Sustainable Textiles: Production, Processing, Manufacturing & Chemistry

Series Editor

Subramanian Senthilkannan Muthu, Chief Sustainability Officer, Green Story Inc., Tung Chung, Hong Kong

This series aims to address all issues related to sustainability through the lifecycles of textiles from manufacturing to consumer behavior through sustainable disposal. Potential topics include but are not limited to: Environmental Footprints of Textile manufacturing; Environmental Life Cycle Assessment of Textile production; Environmental impact models of Textiles and Clothing Supply Chain; Clothing Supply Chain Sustainability; Carbon, energy and water footprints of textile products and in the clothing manufacturing chain; Functional life and reusability of textile products; Biodegradable textile products and the assessment of biodegradability; Waste management in textile industry; Pollution abatement in textile sector; Recycled textile materials and the evaluation of recycling; Consumer behavior in Sustainable Textiles; Eco-design in Clothing & Apparels; Sustainable polymers & fibers in Textiles; Sustainable waste water treatments in Textile manufacturing; Sustainable Textile Chemicals in Textile manufacturing. Innovative fibres, processes, methods and technologies for Sustainable textiles; Development of sustainable, eco-friendly textile products and processes; Environmental standards for textile industry; Modelling of environmental impacts of textile products; Green Chemistry, clean technology and their applications to textiles and clothing sector; Eco-production of Apparels, Energy and Water Efficient textiles. Sustainable Smart textiles & polymers, Sustainable Nano fibers and Textiles; Sustainable Innovations in Textile Chemistry & Manufacturing; Circular Economy, Advances in Sustainable Textiles Manufacturing; Sustainable Luxury & Craftsmanship; Zero Waste Textiles.

Subramanian Senthilkannan Muthu
Editor

Sustainable Textile and Apparel Chain Management

Towards the UN Sustainable Development Goals

Springer

Editor
Subramanian Senthilkannan Muthu
Chief Sustainability Officer
Green Story, Canada
Tung Chung, Hong Kong

ISSN 2662-7108 ISSN 2662-7116 (electronic)
Sustainable Textiles: Production, Processing, Manufacturing & Chemistry
ISBN 978-3-031-80239-3 ISBN 978-3-031-80240-9 (eBook)
https://doi.org/10.1007/978-3-031-80240-9

This Springer imprint is published by the registered company Springer Nature Switzerland AG
The registered company address is: Gewerbestrasse 11, 6330 Cham, Switzerland

This book is dedicated to:
The lotus feet of my beloved Lord
Pazhaniandavar
My beloved late Father
My beloved Mother
My beloved Wife Karpagam and Daughters—
Anu and Karthika
My beloved Brother
Last but not least
To everyone working in various industrial
spaces to make the world "SUSTAINABLE"

Contents

Sustainable Textile and Apparel Chain Management: Towards the UN Sustainable Development Goals

Subramanian Senthilkannan Muthu

The United Nations' Sustainable Development Goals (SDGs) represent a universal call to action to end poverty, protect the planet, and ensure prosperity for all by 2030. These 17 goals address global challenges such as poverty, inequality, climate change, environmental degradation, peace, and justice, forming an interconnected blueprint for a sustainable future. The textile and apparel industry, a significant global economic sector, is both a major contributor to environmental pollution and a vital area for implementing sustainable practices. The textile and fashion industries are at the forefront of the SDGs due to their substantial environmental and social impacts.

They account for 2% of the world's GDP, employ millions of people, and are among the largest polluters, generating over 1.2 billion tons of carbon dioxide emissions annually. The industry's challenges include excessive waste, water pollution from dyeing processes, and the proliferation of microplastics. Despite these issues, the SDGs provide a comprehensive framework to guide the industry towards sustainability, emphasizing responsible production and consumption (SDG 12), climate action (SDG 13), and decent work and economic growth (SDG 8) (Thakker & Sun, 2023). This chapter will present an overview of the entire book.

The second chapter deals with environmental sustainability in textile and apparel global value chain towards achieving the United Nations Sustainable Development Goals. The relocation of textile and apparel manufacturing to the Global South has served as a double-edged sword, spurring economic development while simultaneously contributing to environmental degradation, thus creating a potentially irreparable imbalance that threatens the well-being of future generations in the region. This chapter offers a comprehensive exploration of environmental sustainability within the Textile and Apparel Global Value Chain (T&A GVC), with a particular

S. S. Muthu (✉)
Chief Sustainability Officer, Green Story Inc, Tung Chung, Hong Kong

S. S. Muthu (ed.), *Sustainable Textile and Apparel Chain Management*, Sustainable Textiles: Production, Processing, Manufacturing & Chemistry,
https://doi.org/10.1007/978-3-031-80240-9_1

focus on the critical role played by suppliers in the Global South. By leveraging a novel analytical GVC framework enriched with insights from international business (IB) and environmental justice (EJ) literature, this chapter seeks to address significant gaps in current research, especially the underutilization of these theoretical perspectives in the study of sustainability practices. Four key research gaps are identified: the limited integration of IB and GVC theories in sustainability studies, the underappreciated relevance of EJ principles in GVC analyses, the insufficient focus on the governance role of suppliers in the Global South, and the lack of incorporation of planetary boundaries in sustainability assessments. To bridge these gaps, this chapter introduces an innovative analytical GVC framework, with three extensions, that provides a holistic approach to understanding environmental sustainability within the T&A GVC. This framework facilitates a detailed examination of the six dimensions of GVCs—input-output structures, geographic and industrial reach, governance mechanisms, paths for upgrading, industry stakeholders, and institutional dynamics—emphasizing their interconnectedness and the complexity of promoting sustainability across the value chain.

By advancing a framework that integrates diverse empirical and conceptual studies, this chapter makes a significant contribution to the literature. It challenges the traditional focus on lead firms and advocates for a more inclusive understanding of sustainability that incorporates the perspectives of suppliers from the Global South. The findings underscore the necessity of adopting a holistic approach to GVCs, one that concurrently considers economic, social, and environmental dimensions to achieve sustainable development. The chapter concludes by offering recommendations for future research and practical strategies for stakeholders, aligning environmental sustainability practices with the United Nations Sustainable Development Goals (UN SDGs).

The third chapter deals with the analysis of challenges to the circular economy model of textile supply chain management using toe-tbl approach. In today's world where resource management has gained importance in parallel with the increasing population, although companies prefer to transition to circular economy models, they face several obstacles and cause excessive depletion of resources. Circular economy, which deals with the protection of natural resources, reducing, recycling, and economic recovery of waste, is one of the main issues that should be taken into consideration due to the intensive use of resources in the textile industry. The aim of the study is to investigate possible circular economy challenges in textile supply chain management, considering technology, organization, environment, social, and economic factors (TOE, Technology, Organization, Environment)); TBL, Triple Bottom Line). Within the scope of the research, 22 criteria were classified according to TOE-TBL dimensions and ranked and evaluated using MCDM techniques. In addition, in the results deduced from general challenges, it has been observed that environmental dimensions are at the top of the criteria that companies that will transition to a circular economy model should prioritize.

The fourth chapter discusses the adaptability of UN SDGs by worldwide fashion brands. The fashion industry has always topped in terms of pollution and impact on the globe, which includes emission of greenhouse gases, demanding high water exhausting industry. These aspects make us ask "If fashion is sustainable for the future?" The fashion industry can be more sensible of the issues arising. Sustainability and sustainable development have gained attention all over the globe to avoid over-utilization of resources. The UN has advocated sustainable development goals in the year 2015, to make an assurance that the people will enjoy peace and prosperity by 2030. The new objectives of economic, social, and environmental developments such as environmental protection and economic growth are demonstrated by the SDGs. In this, SDG 12 mentions Responsible consumption and production and this is where fashion brands come in to picture. Due to the increased buying habits of customers, the fashion industry is not able to ascertain the future impacts clearly which leads to irresponsible production and consumption. In this regard, it becomes essential to inculcate the best practices in these brands to be more responsible in production so that the consumption pattern can be changed. Many brands such as Zara, Uniqlo, and Gant are doing many activities in this regard. With the increasing awareness among the people of sustainability, the fashion sector will follow sustainable measures. By acknowledging all these situations there has been relatively little byway of increased regulations specific to the fashion industry to encourage a change in standard practices. Fashion and textile industries require more corporate and socially responsible activities that would propel SDGs. In this chapter, we will discuss the various UN SDG principles followed by fashion brands and other textile industries in India and other countries. The chapter will also throw light on the best practices that can be followed for achieving SDG.

The fifth chapter elaborates on the challenges for the achievement of the Sustainable Development Goals in the South American Andean camelid chain. South American camelids are represented by four species, two domestic species, the alpaca (Vicugna pacos) and the llama (lama glama), and two wild species, the vicuña (Vicugna vicugna) and the guanaco (lama guanicoe), which live in high Andean areas above 3800 metres above sea level. The breeding and production of domestic South American camelids is one of the most important productive and economic activities on which 70 to 80% of the annual family income in the area depends, according to the FAO. In the case of wild species, the "vicuña" is currently categorised in Peru as a "near threatened" species, with more than 50% of the world population in Peru, around 208,899 specimens; and the "guanaco" is categorised as a "critically endangered" species due to its small population of approximately 3810 specimens. The alpaca (Vicugna pacos) is one of the South American camelid species of the Andean biodiversity that has a coat classified as a luxury fibre; considered of high textile value due to its softness, length, fineness, warmth; presenting a range of 22 natural colours and more than 64 intermediate shades. The world population of alpacas exceeds 6 million, 72% of which are concentrated in the Andean region of Peru. However, this activity is not managed under a sustainable model; the production system of domestic South American camelids is extensive grazing, based on the forage supply of natural pastures or rangelands; the degradation of the Andean ecosystem is a complex problem that is related to the organisational model of the communities and land management. Likewise, in animal production and management, a predominantly traditional technology is applied with little adoption of technological innovations; there is the existence of intermediation oriented towards supplying the external market, which generates instability in prices and causes an inequitable distribution along the chain, with producers being the least favoured. There are around 92,700 agricultural and livestock units dedicated to the production and breeding of alpacas and 55,250 llamas, constituting an important source of employment, of which more than 76% are in conditions of poverty and extreme poverty. The implementation of actions with clear targets for the sustainable development of this population of communities and settlers, related to ending poverty (SDG1), ensuring decent work and economic growth (SDG 8), generating spaces for technology transfer to improve the conditions for innovation and infrastructure of the chain (SDG 9) and achieving results in reducing inequalities (SDG 10) are a priority to ensure the subsistence of the South American camelid value chain.

The sixth chapter explains batik and textile supply chain management based on the United Nations' Sustainable Development Goals with the help of Indonesian case studies. In Indonesia, batik is a traditional cloth cultural heritage. Small and medium-sized enterprises (SMEs) are the primary producers of batik cloth. Additionally, there are numerous textile companies which, apart from operating in the field of ordinary commercial fabrics, are also engaged in the production of batik cloth. The aim of this research is to conduct descriptive studies on the batik and textile industries in Indonesia. Indonesia's batik cloth industry is a significant industry. Powered by the UN's Sustainable Development Goals (SDGs) and the significance of sustainability, this study explores ways to align the batik cloth and textile industry in Indonesia's supply chain with the Sustainable Development Goals (SDGs). The method in this study uses literature studies and industrial practices in batik SMEs, we can show that several industrial processes in batik cloth SMEs are in accordance with SDG's. Various industries have begun implementing and trying to reach Sustainable Development Goals (SDGs) such as clean water supply and proper sanitation, equal rights of men and women, good job, and adequate health facilities. Currently, batik cloth SMEs have adopted the use of natural dyes in their dyeing procedures, which are both secure and environmentally friendly. In this study, we propose several perspectives that can be done so that the batik cloth and textile industry in Indonesia can increasingly fulfill the SDGs, as well as the role of stakeholders so that the batik cloth industry in Indonesia becomes more advanced.

The seventh chapter deals with the development and challenges of infant diapers ensuring good health with sustainability aspects: Landscape of UN perspective. The utilization of diapers is a critical component of contemporary infant care, as societal and physiological demands require it of the caregivers in the modern era. This book chapter explores the multi-dimensional world of diapers, explaining their primary functions, diversified varieties, applications and materials. With a discerning eye on current market trends, sustainability imperatives and emerging challenges, the chapter navigates through the intricate landscape of the UN perspective of diaper technology. It also aligns diaper developments with the United Nations Sustainable Development Goals, emphasizing sustainable alternatives that prioritize preserving the environment, ethical consumption, and health. As explored herein, the primary function of diapers extends beyond moisture absorption and liquid retention; it encompasses the containment of waste, maintenance of skin health, and facilitation of mobility. Various diapers, from traditional cloth to modern disposables, are meticulously examined alongside their advantages and limitations. Additionally, this chapter scrutinizes diaper construction and materials, illuminating the intricate engineering and scientific formulations behind their efficacy. Critical attention is devoted to the sustainability criteria governing diaper production and disposal. With scholarly precision, challenges such as toxic materials, frequent skin issues such as diaper dermatitis, and the use of super absorbent polymer (SAP) are dissected. Additionally, the chapter explores the ongoing progress of sustainable solutions to conventional disposable diapers, emphasizing exciting advances and their potential to reduce environmental impact while maintaining efficacy and safety. This chapter also extrapolates potential developments, emphasizing the burgeoning interest in reusable diaper solutions. The study combines scientific investigation with practical insights, making it an authoritative resource for academics, practitioners, and stakeholders committed to promoting baby care via innovation and sustainability.

The eighth chapter explains curbing carbon emissions and its impact on the environment and the steps to achieve UN SDGs. The United Nations Sustainable Development Goals (SDGs) establish an outline for addressing global concerns, including those affecting the textile and garment industry. While crucial to global economies, this industry faces severe sustainability issues that influence multiple SDGs. SDG 12 (Responsible Consumption and Production) emphasises the importance of sustainable practices in the textile and garment industries, such as reducing waste, using resources efficiently, and preventing contamination. Furthermore, SDG 8 (Decent Work and Economic Growth) emphasises the need for fair labour practices and job creation, which are critical in an industry plagued by labour exploitation. In addition, SDG 13 (Climate Action) emphasises the industry's role in lowering carbon emissions and implementing eco-friendly manufacturing practices to reduce environmental damage. The textile and apparel sector may help create a more sustainable future by supporting social justice, economic development, and ecological stewardship through alignment with the SDGs.

The textile and garment industries significantly contribute to carbon emissions, with an enormous environmental impact across their supply chains. Carbon emissions occur at several stages, including the cultivation of raw materials such as cotton or the manufacture of synthetic fibres and the industrial processes involved in spinning, weaving, dyeing, and finishing garments. Furthermore, transporting materials and final products contributes to the carbon footprint. The extensive dependence on fossil fuels for energy in manufacturing facilities worsens the problem. The demand from consumers for fast fashion continues to climb, putting pressure on these companies to address their environmental effects and implement more sustainable practices to reduce carbon emissions. Efforts like using eco-friendly production practices, using renewable energy sources, and fostering circularity in fashion are critical steps towards lowering the carbon footprint of textile and garment manufacturing.

Carbon emissions are tied to the United Nations' Sustainable Development Goals (SDGs). Excessive carbon dioxide and other greenhouse gas emissions exacerbate climate change, making SDG 13: Climate Action a critical challenge. This environmental deterioration, in turn, impedes progress towards other SDGs, including poverty eradication (SDG 1), health and well-being (SDG 3), and sustainable cities and communities (SDG 11). Carbon emission reduction activities are therefore critical in addressing the larger range of sustainable development described by the United Nations. Prioritising climate action allows governments to develop resilience, equity, and prosperity, which aligns with the SDGs' overarching objective of creating a more sustainable future for all.

The last chapter deals with aligning apparel and textile sector practices with the United Nations Sustainable Development Goals and balancing the pillars of sustainability. Uplifted by the United Nations and the pertinence of sustainability, there is a growing awareness of the United Nations Sustainable Development Goals. Primarily, the UN SDGs are concerned with the three pillars of sustainability. The pillars include social, economic, and environmental concerns. The preceding research on these pillars has been majorly focused on any one of these pillars leaving other aspects with insufficient data explorations. This chapter will critically analyse the literature and the existing industrial practices and discuss the correlation between the apparel and textile sectors and the UN SDGs. It also studies how the three pillars of sustainability in UN SDGs are handled and achieved in apparel and textile supply chains. It examines strategies for effectively balancing economic growth, environmental protection, and social equity, recognising the interconnectedness of these elements in driving sustainable development. It is known that every organisation or sector is equally responsible for the activities towards development in these three aspects. The majority of the existing industrial practices show that the present sustainable practices are not fulfilling the objectives of the SDGs as stated by the UN. Many types of actions can be taken by the apparel and fashion industry stakeholders to balance the pillars. Stakeholders' sustainable development plans and actions concerned with the sustainability pillars and UN SDGs are proposed. The future research agenda and the critical areas of scope are also highlighted.

Acknowledgment I take this opportunity to thank all the contributors for their earnest efforts to bring out this book successfully. I am sure readers of this book will find it very useful.

With best wishes,

Dr. Subramanian Senthilkannan Muthu

Reference

Thakker, A. M., & Sun, D. (2023). Sustainable development goals for textiles and fashion. *Environmental Science and Pollution Research, 30*, 101989–102009. https://doi.org/10.1007/s11356-023-29453

Environmental Sustainability in Textile and Apparel Global Value Chain: Towards Achieving the United Nations Sustainable Development Goals

Mohammad Mobarak Hossain, Omar Al-Tabbaa, and Mohammad Faisal Ahammad

1 Introduction

This chapter is dedicated to synthesizing and enhancing our current understanding of the environmental sustainability within the textile and apparel global value chain (T&A GVC), with a pronounced focus on suppliers in the Global South. We explore sustainable development in this sector, helping stakeholders like practitioners, policymakers, and consumers recognize their roles in promoting sustainability. We introduce a novel analytical GVC framework with three extensions to comprehensively address the nuances within the T&A GVC, particularly focusing on environmental sustainability from the perspective of suppliers in the Global South.

Recent decades have witnessed transformative shifts in the T&A industry due to technological advancements, the rise of emerging economies, and eased trade barriers, significantly altering the landscape of multinational enterprises (MNEs) (Gereffi & Lee, 2012; Buckley & Strange, 2015; Kano, 2018). The shift of manufacturing bases from the Global North to the Global South has notably turned this sector into a complex GVC, catalyzing economic growth and potentially lifting regions out of poverty despite causing serious environmental degradation (Pahl & Timmer, 2020; Lopez-Acevedo & Robertson, 2016; Hussain et al., 2020; Hollweg, 2019; Buckley & Strange, 2015; Khan et al., 2019; Alves et al., 2022; Boström & Micheletti, 2016; Khattak et al., 2015; Li et al., 2022).

The GVC framework, enriched by inputs from International Business (IB) and environmental justice (EJ) literature, serves as a theoretical lens to evaluate

M. M. Hossain (✉) · O. Al-Tabbaa · M. F. Ahammad
International Business Department, Leeds University Business School (LUBS), University of Leeds, Leeds, UK
e-mail: bnmmh@leeds.ac.uk; o.f.o.altabbaa@leeds.ac.uk; m.f.ahammad@leeds.ac.uk

S. S. Muthu (ed.), *Sustainable Textile and Apparel Chain Management*, Sustainable Textiles: Production, Processing, Manufacturing & Chemistry,
https://doi.org/10.1007/978-3-031-80240-9_2

economic, social, and environmental dimensions within this sector. Scholars have applied this framework to analyze how evolving dynamics influence sustainability, with a noted focus on economic and social upgrades over environmental concerns (Milberg & Winkler, 2013; Golgeci et al., 2021; Krishnan et al., 2023; Khan et al., 2019; De Marchi et al., 2019).

In this context, the environmental trajectory of the GVC has gained increasing relevance, driven by consumer awareness, regulatory demands, and civil activism, urging firms to adopt sustainable strategies for competitive advantage (Porter & Kramer, 2006; Orsato, 2006; Poulsen et al., 2016; De Marchi et al., 2019; Gereffi, 2019; Gölgeci et al., 2019; Navarrete et al., 2020; Krauss & Krishnan, 2021). Despite this growing interest, our review identifies four critical gaps in how environmental sustainability is currently studied within the textile and apparel global value chain.

Firstly, there is a marked lack of comprehensive application of IB and GVC theories in analyzing how stakeholders within the T&A GVC, especially from the Global South, collaborate to enhance environmental sustainability (cf., De Marchi et al., 2014; De Marchi et al., 2020; Sinkovics et al., 2017; Sinkovics & Sinkovics, 2019; Khan et al., 2019; Langford et al., 2023). This gap suggests a need for more integrative research approaches that combine these theoretical frameworks. Secondly, the relevance of environmental justice (EJ) literature—which advocates for the fair treatment and meaningful involvement of all people in environmental laws, regulations, and practices (EPA, 2023)—in GVC studies is often underestimated. This oversight includes a failure to consider how EJ principles—distributive,[1] procedural,[2] and recognition-based justice[3]—impact suppliers' experiences and operations within the T&A GVC (Waite, 2014). Integrating EJ perspectives could provide deeper insights into equitable sustainable development.

Thirdly, insights into the multifaceted roles of suppliers from the Global South in governing environmental sustainability within GVCs are lacking. Current literature does not sufficiently explore how these suppliers contribute to sustainability governance (Gereffi & Lee, 2012; Achabou et al., 2017; Kano, 2018; Golgeci et al., 2021; Langford et al., 2023; Khattak & Stringer, 2016; Ponte et al., 2023). Finally, there is a critical need for research that incorporates planetary boundaries into environmental sustainability assessments in the GVC context. The existing metrics for evaluating sustainability are often inadequate in reflecting true environmental costs and benefits across different sectors and regions (De Marchi & Gereffi, 2023; Krishnan et al., 2023; De Marchi et al., 2019).

[1]Distributional justice entails the fair distribution of environmental costs and benefits among the stakeholders.

[2]Procedural justice focuses on fairness in decision-making processes and how individual and social groups are included.

[3]Recognition justice focuses on how the varied experiences and vulnerabilities of diverse groups are recognized.

Consequently, this chapter synthesizes existing literature on T&A industry and elucidates environmental sustainability within the T&A GVC, with a special emphasis on suppliers from the Global South. This is important because existing practices of environmental sustainability have failed to achieve their intended goals, as a result, the T&A industry remains the second-largest polluting industry worldwide (Gereffi & Frederick, 2010; BGMEA, 2020; Niinimäki et al., 2020). In this study, we explore a series of interrelated research questions focused on the environmental sustainability within the T&A GVC of the Global South. First, we investigate how various stakeholders influence the environmental sustainability processes of suppliers from the Global South. Next, we examine the ways in which these suppliers navigate and implement environmental sustainability processes. Finally, we assess how they evaluate the environmental performance of these initiatives. Our aim is to enrich the discourse on sustainable development within T&A GVC.

To achieve this, we introduce a pioneering analytical GVC framework along with three supplementary extensions. This approach systematically illustrates the application of the GVC framework in comprehending and improving the environmental sustainability practices of suppliers in the Global South within the T&A GVC. The framework is designed to inform the best practices for a wide range of stakeholders, spanning from the Global South to the Global North, promising to shape a more sustainable future in the industry. More specifically, we propose that the six dimensions of the existing GVC framework—input-output structures, geographic and industry reach, governance mechanisms, paths for upgrading, industry stakeholders, and institutional dynamics (Fernandez-Stark et al., 2013; De Marchi et al., 2020)—can shed light on various aspects of environmental sustainability within the T&A GVC (De Marchi & Gereffi, 2023; Krishnan et al., 2023). For example, input-output analysis can identify production stages heavily implicated in environmental degradation. Governance mechanisms can elucidate the forces that motivate suppliers from the Global South to engage in environmental sustainability process (ESP), while pathways for upgrading offer a framework to evaluate the effectiveness of ESPs.

Our research makes three significant contributions to the literature on global value chains (GVC), international business, and environmental justice. Firstly, we have developed an analytical GVC framework that synthesizes diverse empirical and conceptual studies related to the environmental sustainability of suppliers from the Global South. This framework not only advances our understanding but also guides future research in environmental sustainability across different GVCs. Secondly, our study challenges the traditional focus of existing literature that typically examines the role of lead firms using GVC governance dimensions and supplier participation through the paths of upgrading within GVCs. We demonstrate the importance of considering all dimensions concurrently when exploring environmental sustainability for suppliers from the Global South. Our findings reveal that a holistic approach is essential for truly understanding and enhancing the mutual benefits of participation in GVCs, thereby ensuring comprehensive economic, social, and environmental advancements. Lastly, our research enhances understanding of environmental sustainability in the Global South by identifying critical gaps in current environmental governance that fail to address distributional, procedural, and recognition-based injustices.

This chapter is organized as follows: It begins with a concise overview of the shift in T&A manufacturing from the Global North to the Global South. Next, we introduce a novel analytical framework based on six GVC dimensions designed to aid in synthesizing the literature on environmental sustainability within the T&A industry. Utilizing this framework, we review and synthesize the relevant literature to address the three research questions previously outlined. Additionally, we examine the connection between environmental sustainability in the T&A GVC and various United Nations Sustainable Development Goals. The chapter concludes with a summary and recommendations for future research.

2 Transition of T&A Production from Global North to Global South

The literature highlights the shift of manufacturing, including T&A manufacturing, from the Global North to the Global South, initiated in the 1970s due to rising production costs in developed countries and the forces of globalization. This shift was propelled by the Global South's lower labor costs, making it a major hub for T&A manufacturing (Divrik & Baykal, 2022; Buckley & Strange, 2015; Khan et al., 2009; Kim et al., 2006; Smith, 1996). In addition, the move of extractive and highly polluting industries to regions with lower production standards and labor costs has been a notable trend, driven by increasing awareness of environmental and social impacts (Reed, 2002; Mohai et al., 2009).

Moreover, a pivotal moment was the end of the quota-restricted Multi-Fiber Arrangement in 2005, which reshaped the global textile and apparel value chains, with China becoming a dominant player (Kar & Kar, 2015; Zhang & Hathcote, 2008). Post-2007/08 financial crisis, China's role in driving the textile industry eastward became even more pronounced (Bressán, 2022; Xu et al., 2018).

Figure 1 visually summarizes this dynamic, detailing the roles of various GVC actors, the flows of information and products, and the involvement of entities from both the Global North and Global South at different stages of the T&A value chain.

Figure 1 delineates the T&A GVC, starting with the comprehensive *production stage*, which is segmented into three phases and displays both upstream and downstream activities, as well as the intermediary roles that connect them.

Raw Materials Production (Upstream):
At the outset of the GVC, this phase includes the cultivation of seeds and processing of fibers, providing the essential raw materials, both natural and synthetic, which are the building blocks for textile production.

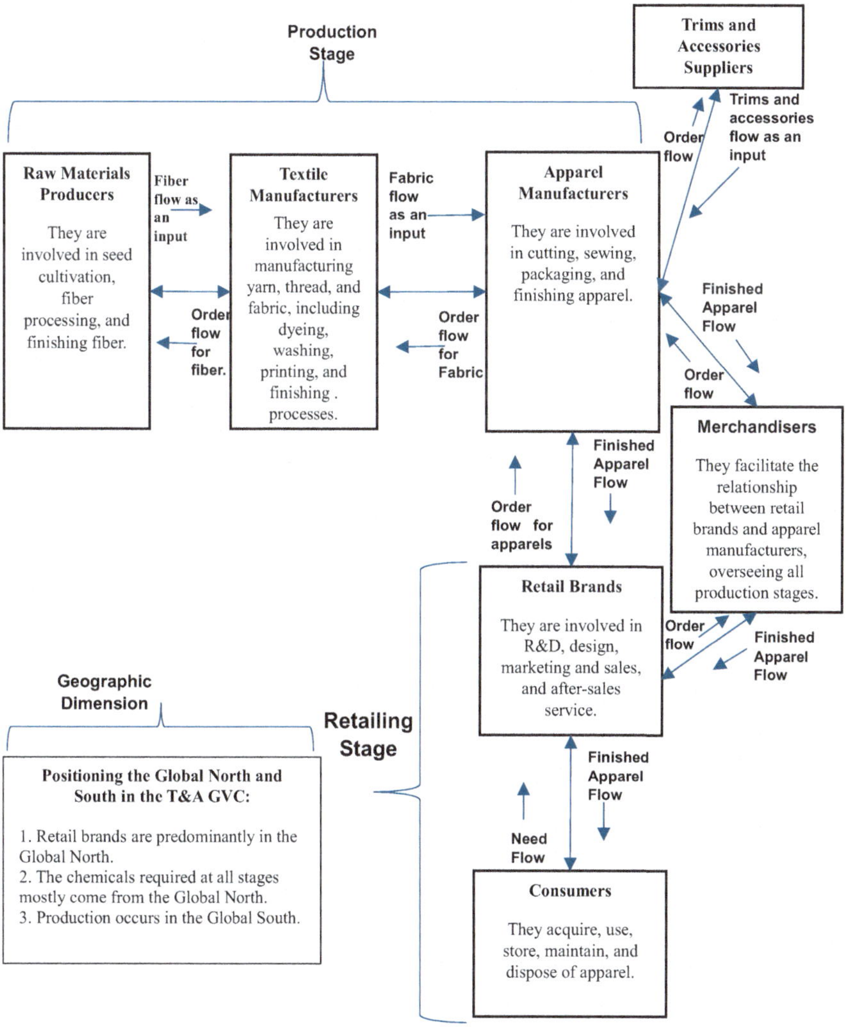

Fig. 1 The main segments of the textile and apparel global value chain. (Source: Adapted from Fernandez-Stark et al. (2011), Connell and Kozar (2014), Laitala et al. (2014), Winakor (1969), Duke University (n.d.))

Textile Production (Midstream):
Here, raw fibers are spun into yarn and subsequently knitted or woven into fabric. The textiles are then refined with additional processes such as dyeing, washing, and printing. This stage stands between upstream raw material production and downstream apparel manufacturing, transitioning the goods downstream.

Apparel Production (Downstream):
This final phase within the production stage is where the apparel is crafted from the textiles. The process involves cutting, sewing, packaging, and adding the finishing touches. Once the apparel is ready, it can be passed along to the retail brands either directly or through the involvement of merchandisers.

Merchandisers as Intermediaries:
Merchandisers function as pivotal intermediaries in the T&A GVC. Retail brands may choose to place orders directly with apparel manufacturers or may opt to work through merchandisers. When engaged, merchandisers take on the responsibility of ensuring that apparel manufacturers and other actors in the chain adhere to the retail brands' specifications and timelines. They oversee the entire process, from the production stage to the delivery of finished apparel, effectively bridging the gap between retail brands and manufacturers. Their oversight is critical in maintaining quality control, meeting deadlines, and facilitating communication across the value chain.

Retailing Stage and Downstream to Upstream Feedback:
The *retailing stage* involves activities by retail brands such as research and development (R&D), design, marketing, and sales, along with after-sale services. This stage is intricately connected to the final consumers, who interact with the apparel products. Consumer feedback and market trends travel upstream, influencing decisions at the merchandising, apparel production, and textile production stages.

Geographic Dimension:
The *geographic dimension* of the chart emphasizes the Global North's predominance in retailing and chemical supply, while the Global South is the main hub for production activities. Merchandisers, often situated within or closely aligned with the markets of the Global North, are integral to ensuring that the production activities in the Global South align with the dynamic needs of the retail brands and the end consumers. This complex web of interactions illustrates the multifaceted nature of the T&A GVC, with merchandisers playing a key role in maintaining the integrity and responsiveness of the entire chain.

Vertical Integration Versus Multiple Firms:
In some instances, the production stages and merchandising activities are vertically integrated within a single firm, allowing for greater control over the supply chain, consistency in quality, and a streamlined process from raw materials to finished products.

Alternatively, the GVC may involve multiple specialized firms. Each firm performs distinct functions across the various stages, providing flexibility and specialization that can drive innovation and efficiency. This multi-firm structure requires careful coordination to maintain the flow of materials, information, and products throughout the value chain.

The chart encapsulates not just the flow of goods from raw materials to consumers but also the potential organizational structures that govern these processes. It reflects the diversity of approaches, from vertical integration, which can minimize costs and ensure consistency, to a more fragmented structure with multiple firms, which can foster specialization and adaptability. Both approaches aim to serve the dynamic needs of Global North's retail brands and consumers, while leveraging the production capabilities of Global South.

3 Environmental Sustainability Within the T&A GVC

3.1 Introduction

This chapter delves into the environmental sustainability challenges and practices within the T&A GVC, placing a significant emphasis on the role and experiences of the Global South. Through a comprehensive review and analysis of existing literature across international business (IB), global value chains (GVC), and environmental justice (EJ) domains, we aim to shed light on how suppliers in the Global South navigate sustainability within the T&A GVC. A key component of our approach is the introduction of an innovative analytical GVC framework, complete with three extensions, designed specifically to dissect and understand these complexities. This framework serves as the cornerstone of our analysis, enabling a deep dive into the six dimensions of the GVC framework to enhance our understanding and strategies for improving environmental sustainability, especially from the perspective of the Global South. By aligning these practices with the United Nations Sustainable Development Goals (UN SDGs), we highlight avenues for collaborative advancement. The insights gained from applying this analytical framework lay the groundwork for future research directions aimed at leveraging the GVC framework to promote environmental sustainability in the T&A sector, demonstrating the entirety of our planned investigation.

3.2 GVC Analytical Framework for Environmental Sustainability Within the T&A GVC

GVCs represent the intricate network of processes involved in producing and delivering a product or service across various countries, each contributing uniquely to the value added (Gereffi, 2005; Lopez-Acevedo & Robertson, 2016). The literature on GVC Framework delves into how "lead firms" orchestrate global production by externalizing certain activities within the value chain, with these pivotal firms originating from both developed and emerging countries (Buckley & Strange, 2015; De Marchi et al., 2020). This framework, characterized by six analytical dimensions, as depicted in Fig. 2, offers a comprehensive tool for assessing the environmental

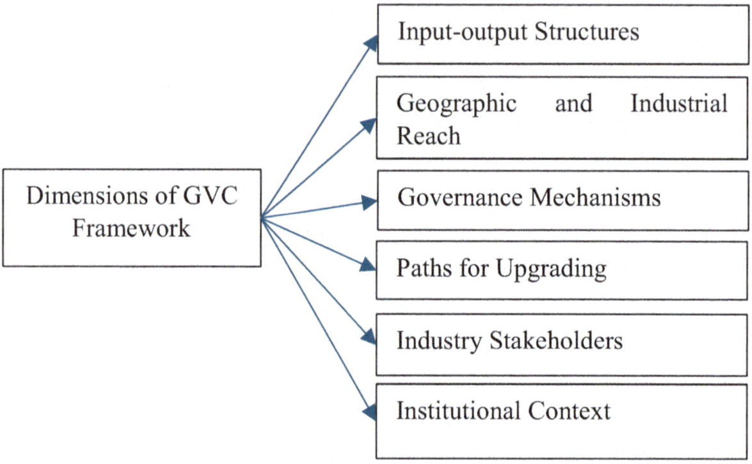

Fig. 2 Six dimensions of GVC Framework. (Source: adapted from Fernandez-Stark and Gereffi (2019), De Marchi et al. (2020))

sustainability within the GVC including T&A (Gereffi et al., 2005; Fernandez-Stark et al., 2013; De Marchi et al., 2020; De Marchi & Gereffi, 2023).

The GVC framework examines the global economy through two distinct perspectives: the "top-down," which centers on the governance of GVCs by lead firms and the structuring of international industries, and the "bottom-up," which emphasizes the upgrading strategies employed by countries, regions, and other stakeholders to enhance or sustain their roles in the global economy (Fernandez-Stark & Gereffi, 2019). However, this division into top-down and bottom-up perspectives fails to capture the essence of collaboration and the role of suppliers in environmental sustainability within the T&A GVC. Therefore, we intentionally avoided this division in Fig. 2.

It is notable to mention that the concept of sustainability (i.e., economic, social, and environmental sustainability) in GVC literature is often discussed in terms of path of upgrading, including economic, environmental, and social upgrading. Table 1 offers a comparative analysis of the three types of upgrading pathways as discussed in the GVC literature.

In exploring the six dimensions of the GVC framework, it becomes evident that these dimensions are not only interconnected but also reinforce one another. For example, pathways for economic upgrading play a crucial role in promoting economic growth while simultaneously shedding light on strategies to strengthen the other two sustainability pillars—social and environmental (Golini et al., 2018). Such pathways also guide us in selecting appropriate ESP to navigate the trade-offs inherent in the triple bottom line of sustainability, which includes economic, social, and environmental considerations (Nand et al., 2022).

By understanding and leveraging these intersections, we have deliberately aligned the six dimensions of the GVC framework to facilitate a more focused

Table 1 Comparison of economic, social, and environmental upgrading in GVCs

Aspects	Economic upgrading	Social upgrading	Environmental upgrading
Definition	The process of moving up to higher value-added activities in the production process or global value chain (Gereffi et al., 2005)	The process of improving the rights and entitlements of workers and enhancing the quality of their employment (Barrientos et al., 2011)	The process of improving environmental performance within the value chain (Jeppesen & Hansen, 2004)
Key focus	Product development, process innovation, functional upgrading, intersectoral upgrading (Humphrey & Schmitz, 2002)	Working conditions, worker rights, wage levels, social protection (Rossi, 2019)	Sustainable resource use, waste management, carbon footprint reduction, compliance with environmental standards (De Marchi et al., 2019)
Importance	Contributes to higher value-added activities and long-term sustainability, increased profits, market share, and competitive advantage (Schmitz, 2006)	Ensures fair and decent working conditions for workers, improved worker satisfaction, reduced turnover, enhanced company reputation (Barrientos et al., 2011)	Reduced environmental impact, compliance with regulations, improved company reputation, potential for cost savings (Khan et al., 2019)
Challenges	Requires overcoming barriers to entry and competition in global markets (Humphrey & Schmitz, 2002)	Faces tensions between meeting social and economic upgrading (Milberg & Winkler, 2011, Nand et al., 2022)	Requires balancing economic with environmental upgrading (Nand et al., 2022, De Marchi, Di Maria, & Ponte, 2013b)

Source: Created by the authors

understanding of environmental sustainability within GVCs including T&A. Figure 3 introduces the analytical GVC framework designed to aid in our analysis of the existing literature to understand how the environmental sustainability occurs within the T&A GVC.

Figure 3 delineates a novel analytical framework for understanding the intricate mechanisms of environmental sustainability in the T&A GVC. At the core of this framework are the drivers (i.e., input for ESP) that motivate or compel suppliers to engage in the environmental sustainability process (ESP) (i.e., process for implementing environmental upgrading strategies), which, in turn, is expected to improve the environmental performance (EP) (i.e., output from ESP) of the suppliers throughout the GVC.

The drivers of ESP are varied; they may include regulatory pressures, consumer demand, competitive advantage, or ethical commitments. They serve as the catalysts that prompt suppliers to adopt sustainable practices. As a result of these drivers, the ESP unfolds as a series of actions and strategies implemented to enhance sustainability, leading to improved EP, which reflects the tangible outcomes of these efforts such as reduced waste, lower emissions, and better resource management.

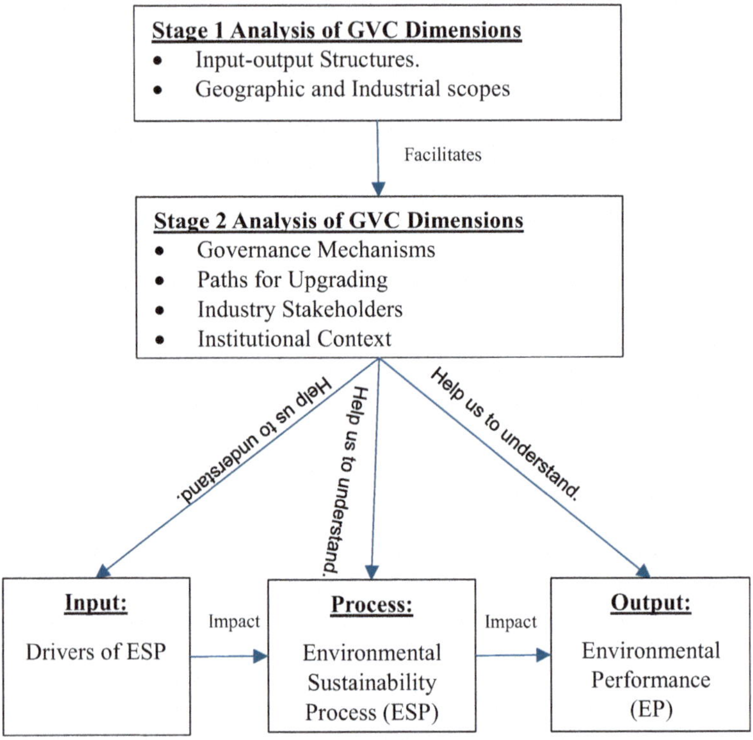

Fig. 3 GVC analytical framework for environmental sustainability within the T&A GVC. (Source: Created by the authors)

However, to fully understand the dynamics at play, one must probe further: Who exactly are these drivers and what are their influences and motivations? How do they influence? How does the ESP manifest within the GVC and through what actions and policies? And importantly, how do we measure the EP to ascertain the real impact of these sustainability efforts?

To answer these critical questions, the figure presents a two-stage analysis based on six dimensions of the GVC framework:

Stage 1 analysis of GVC dimensions focuses on the foundational aspects of the value chain, examining the *input-output structures* that detail the flow of information, materials, and services, as well as the *geographic and industrial scopes* that establish the breadth and reach of the chain's operations.

Stage 2 analysis of GVC dimensions advances the inquiry by looking into the more nuanced and operational aspects of the GVC. This includes understanding the *governance mechanisms* that dictate how sustainability is managed within the chain, the *paths for upgrading* that highlight how suppliers can improve their environmental impact, the *roles of industry stakeholders* in influencing and supporting ESP, and the *institutional context* that frames the regulatory and normative environment in which suppliers operate.

Together, these stages of analysis serve as a comprehensive tool for unraveling the complexities of environmental sustainability within the T&A GVC. They allow for an in-depth examination of each component, shedding light on the "who," "how," and "why" of the drivers of sustainability, the processes involved in ESP, and the metrics for assessing EP. This systematic approach provides stakeholders with the insights necessary to effectively guide and improve sustainability practices across the global value chain. Below, we will explore and synthesize the literature on the T&A industry by applying the analytical GVC framework outlined above.

3.2.1 Drivers for Environmental Sustainability Process Within T&A GVC

Research on the drivers of ESP in GVCs is divided into two main perspectives. One highlights external influencers such as lead firms, state policies, consumers, and NGOs, which are seen as primary catalysts for ESP adoption (Jeppesen & Hansen, 2004; De Marchi et al., 2019). Lead firms often set environmental benchmarks in GVCs, significantly influencing ESP adoption across supply levels (Khattak & Stringer, 2016; Reps & Braun, 2012; Achabou et al., 2017; Buckley & Strange, 2015; Van Assche & Narula, 2023). Similarly, government policies can play a vital role in promoting or mandating ESP (De Marchi & Alford, 2022; Diabat et al., 2014; Zhou et al., 2022). On the internal side, competitive advantage is a significant motivator for adopting ESP, with internal stakeholders like managers and employees actively supporting these initiatives (Hoang et al., 2022; Khattak et al., 2015).

However, there is no consensus in the literature about the primary drivers of ESP in various industries and value chain stages. For example, while the T&A industry mainly responds to external pressures such as NGO advocacy, the maritime industry focuses on internal cost reductions and efficiency improvements to drive sustainability, often in the face of insufficient external regulations (De Marchi et al., 2019). In addition, final consumers in the textiles and apparel (T&A) industry do not directly influence suppliers; instead, they affect the practices of retail brands (Nguyen et al., 2014).

Figure 1 illustrates the potential direct and indirect influences among various actors, highlighting how multinational enterprises (MNEs) indirectly attempt to control sub-suppliers through suppliers, although this strategy often fails to achieve the desired results (Van Assche & Narula, 2023). Building on this analysis, Fig. 4 synthesizes the primary drivers of environmental sustainability within each stage of the value chain. To fully understand these dynamics, it is imperative to integrate insights from the GVC framework, international business literature, and stakeholder theory, as demonstrated by Nguyen et al. (2014).

This intricate landscape of drivers, both external and internal, and their varying impact across industries, underscores the complexity of pinpointing definitive

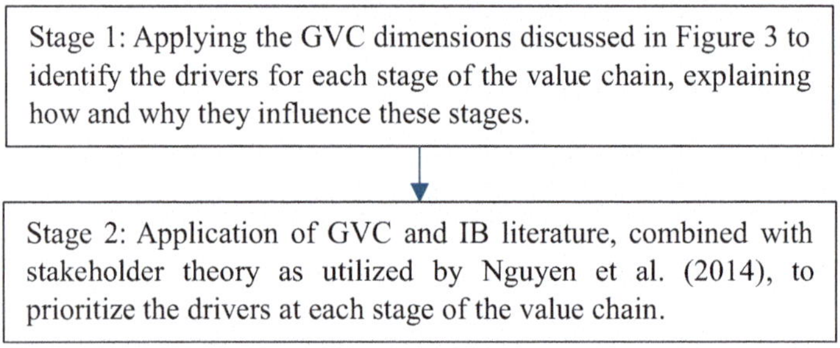

Fig. 4 GVC analytical framework for identifying drivers of ESP within the T&A GVC (an extension of Fig. 3). (Source: Created by the authors)

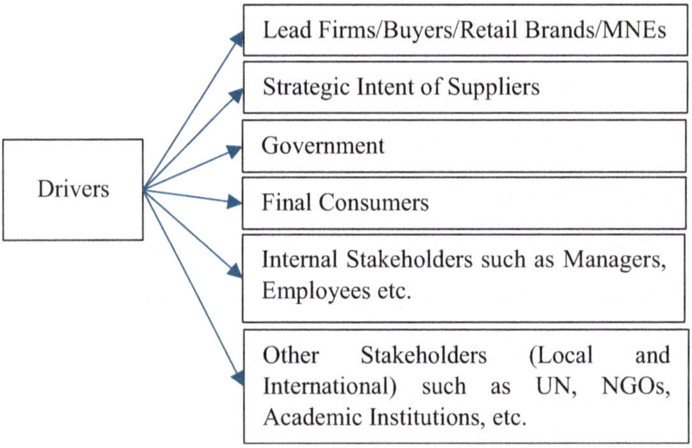

Fig. 5 Drivers for ESP within the T&A GVC. (Source: Created by the authors)

motivators for ESP adoption in T&A GVC. To visually encapsulate this complexity, Fig. 5 delineates the key entities instrumental in fostering the ESP within the T&A GVC.

Below, let us briefly discuss the roles of key stakeholders who drive ESP of suppliers from the Global South within the T&A GVC.

3.2.1.1 Lead Firms/Buyers/Retail Brands/MNEs

Lead firms, often referred to as buyers, retail brands, or multinational enterprises (MNEs) in the T&A GVCs, wield substantial market power that is crucial for driving ESP through strategic governance. These firms are the main catalyst (Nguyen et al., 2014), who set environmental standards that their suppliers, especially those in the Global South, must follow to maintain or improve their competitive standing within the market (De Marchi & Di Maria, 2019; Buckley & Strange, 2015; Khattak et al., 2015).

The motivation for lead firms to promote sustainability is multi-faceted. Consumer demand for eco-friendly products is growing, which pushes these companies to implement greener practices to sustain their market position and enhance their brand prestige (Ponte, 2021; Diabat et al., 2014). Additionally, stringent environmental regulations and the potential for increased costs due to non-compliance compel these firms to adopt more sustainable operations (Diabat et al., 2014). Moreover, there is an economic incentive as sustainable practices often lead to reduced costs and increased operational efficiency, such as lower energy consumption and waste reduction (Meng et al., 2022; Jensen & Whitfield, 2022).

To achieve these goals, MNEs employ a variety of strategies that balance economic and environmental objectives. These strategies include rigorous selection of suppliers who adhere to high environmental standards, partnerships with non-profits organization, and the implementation of strict sustainability policies to meet consumer expectations and regulatory requirements (Ponte, 2021; Li et al., 2022; Meng et al., 2022; Li et al., 2024; Kano, 2018; Golgeci et al., 2021; Uddin et al., 2023; Lee, 2016; Van Assche & Narula, 2023).

Notable companies like H&M, Patagonia, and The North Face are leading the way in integrating sustainability into their business models by employing sustainable product strategies, investments, performance evaluations, corporate social responsibility initiatives, and environmental management systems (Shen et al., 2017). These firms ensure compliance through codes of conduct (COC) and audit-based monitoring systems, complemented by eco-labeling and certification programs that not only boost supplier reputation but also align long-term profitability with sustainability (Shen et al., 2017; Jeppesen & Hansen, 2004; De Marchi et al., 2019).

Additionally, lead firms foster an innovative environment conducive to eco-friendly processes. By prioritizing environmental integrity alongside business interests, they not only promote compliance but also support continuous learning and collaboration among suppliers. This holistic approach significantly enhances the suppliers' ability to meet rigorous environmental standards and contributes to the overall sustainability of the GVC (Lister et al., 2015; Golini et al., 2018; Cheung et al., 2010; Peng et al., 2022).

In conclusion, the role of lead firms in promoting ESP within T&A GVCs is crucial and multi-dimensional. Through stringent standards, strategic partnerships, and a balanced focus on ecological and economic sustainability, these firms are vital catalysts in driving the industry toward more sustainable practices. Their efforts ensure that environmental sustainability is woven into the fabric of global textile and apparel value chains, marking significant progress in the industry's journey toward greener operations.

3.2.1.2 Strategic Intent of Suppliers

Suppliers from the Global South within T&A GVCs are strategically adopting ESP to secure a competitive advantage and enhance their environmental credentials. This strategic intent is multifaceted, driven by the desire to enhance their reputation, benefit from knowledge spillovers, forge stronger relationships with lead firms, target eco-conscious consumers, optimize resource efficiency, and penetrate new markets (De Marchi et al. 2013a, b; Khattak & Park, 2018).

Key strategies employed include improving operational efficiency and reducing energy consumption, which not only contribute to environmental sustainability but also bolster competitive positioning (Khattak et al., 2015). Moreover, aligning with the environmental standards of lead firms not only opens up new market opportunities but also enhances overall competitiveness. This strategic alignment facilitates the adoption of new technologies and best practices, significantly boosting environmental performance. A prime example is VIYELLATEX in Bangladesh, which has attracted buyer preference through efficient resource use and strict adherence to sustainability standards (Khattak & Park, 2018).

Additionally, by employing dynamic capabilities to navigate environmental challenges, suppliers from the Global South are enhancing their sustainability profiles (Khattak & Stringer, 2016; Golgeci et al., 2021; Krishnan et al., 2023; Uddin et al., 2023). The integration of sustainability into financial strategies is also crucial, as it underpins long-term viability and competitiveness (Akbar & Ahsan, 2020; De Marchi, Di Maria, & Micelli, 2013a).

Overall, the strategic intent of suppliers in T&A GVCs, particularly those from the Global South, is to achieve competitive advantage through operational efficiency, market exploration, and rigorous environmental standards. These efforts illustrate a comprehensive and strategic approach to environmental sustainability, aligning with the broader objectives of sustainable global trade and demonstrating a profound commitment to ecological stewardship.

3.2.1.3 Government

Government intervention plays a crucial role in advancing ESP within T&A GVCs. Research, including recent studies by Ponte et al. (2023), highlights the significant impact of horizontal governance—such as government policies and regulations—on promoting sustainability in GVCs. This form of governance is increasingly seen as essential, not only due to the high pollution potential of the T&A industry but also because of the need for specialized knowledge, skills, and commitment to drive the industry toward greater sustainability (Ghosh & Mohan, 2021).

Governments facilitate this transformation through a variety of mechanisms. Environmental regulations, infrastructure policies, and financial incentives such as tax breaks and subsidies help industries comply with regulatory standards and meet consumer demands for eco-friendly products (De Marchi & Alford, 2022; Khan et al., 2019; Nguyen et al., 2014; Van Assche et al., 2023). These measures are designed to encourage firms within the GVC to adopt more sustainable practices by making it economically viable and strategically beneficial to do so.

Additionally, governments play a key role in managing resource use and promoting sustainable industrial growth. Studies on the Chinese textile industry and BRIC countries show that targeted governmental strategies in these areas can significantly influence environmental outcomes (Zhou et al., 2022; Guan & Qamruzzaman, 2022; Mahmood et al., 2021). Public procurement policies, economic directives, and comprehensive regulatory frameworks further shape the strategic decisions of firms in GVCs, often prompting them to adjust their business strategies to align with governmental goals, especially in hopes of securing public contracts (Van Assche et al., 2023).

Overall, the effectiveness of government intervention in the T&A sector underscores the importance of coordinated governance and detailed policies. By setting clear guidelines and providing necessary incentives, governments are pivotal in underpinning environmental stewardship and promoting sustainable development throughout the global value chains. This comprehensive approach ensures that firms not only strive to meet immediate compliance needs but also work toward long-term environmental sustainability.

3.2.1.4 Final Consumers

Final consumers play a pivotal role in driving ESP among suppliers within the Global South's T&A GVCs. Their growing demand for sustainable products acts as a powerful catalyst for change, encouraging suppliers to implement green practices despite often noted discrepancies between consumers' stated environmental concerns and their actual purchasing behaviors (Jegethesan et al., 2012; Nguyen et al., 2014; De Marchi et al., 2019; Das, 2021).

Consumers enhance environmental sustainability by increasingly becoming aware of the ecological impacts of their apparel choices. This heightened awareness is largely shaped by innovative communication strategies—such as detailed hangtags and the use of digital platforms—that inform and influence consumer behavior (Brosdahl & Carpenter, 2010; Das, 2021; Li & Leonas, 2021; Mastamet-Mason & More, 2023). Additionally, the provision of transparent information and clear labeling, including standardized symbols and comprehensive product descriptions, plays a crucial role in helping consumers make environmentally conscious decisions (Filho, 2012).

In conclusion, the influence of final consumers is critical in promoting ESP within the T&A sector of the Global South. Their preference for sustainability, driven by an increased awareness and demand for transparency and clear labeling, motivates suppliers to adopt and continuously improve their green practices. This consumer-driven pressure not only fosters individual supplier compliance but also catalyzes broader industry shifts toward enhanced environmental performance, underscoring the significant impact of consumer behavior on sustainability in global value chains.

3.2.1.5 Internal Stakeholders

The internal stakeholders—employees, managers, and investors—play a significant role in driving ESP within GVCs of the T&A industry. Each group contributes uniquely to the promotion of environmental sustainability, creating a synergistic effect that fosters widespread adoption of ESP.

Employees, particularly those affected by occupational pollution, are key proponents of sustainability. Their firsthand experiences with pollution drive them to advocate for initiatives aimed at reducing environmental impact, thereby catalyzing the adoption of ESP within their organizations (Nguyen et al., 2014; Diabat & Govindan, 2011). This grassroots advocacy is crucial as it highlights the direct benefits of sustainability from within the workforce.

Managers amplify this momentum by integrating environmental benchmarks into the operational strategies of their firms (Nguyen et al., 2014). They not only set these standards but also implement reward systems for suppliers who consistently meet these criteria, thus encouraging a shift toward more sustainable practices. Their leadership in prioritizing ecological concerns helps to align business practices with environmental sustainability, focusing on innovative products and sustainable economic growth (Hoang et al., 2022; Jia et al., 2020; Diabat & Govindan, 2011).

Investors further reinforce this trend by channeling funds toward firms that demonstrate robust environmental policies and practices. This investment strategy motivates suppliers to adhere to stringent environmental standards, knowing that compliance can attract crucial financial support. Investors' focus on sustainable value chain management practices emphasizes the financial viability of adopting ESP, making it an attractive strategy for forward-thinking businesses (Hoang et al., 2022; Jia et al., 2020; Koltonuk, 2020).

In conclusion, the combined efforts of employees, managers, and investors are instrumental in promoting environmental sustainability within the T&A GVCs. By advocating for pollution reduction, setting clear environmental objectives, rewarding compliance, and financially supporting environmentally responsible firms, these internal stakeholders collectively drive the green transition. Their actions ensure that ESP is not only a key element of corporate social responsibility but also a central component of operational and strategic decision-making in the industry.

3.2.1.6 Other Stakeholders (Local and International)

A diverse array of local and international stakeholders significantly influences ESP among suppliers, particularly in the Global South, within T&A GVCs. These stakeholders range from international organizations such as the United Nations Environment Program (UNEP), International Labor Organization (ILO), European Union (EU), and World Trade Organization (WTO), to Civil Society Organizations (CSOs), local communities, environmental advocacy groups, academic institutions, and industry associations. Each plays a unique role in shaping sustainable practices across the industry.

Organizations like UNEP, ILO, EU, and WTO are pivotal as they establish critical environmental and social standards and enforce stringent regulations. By setting these benchmarks, they ensure that suppliers in the GVC must adopt sustainable and ethical practices to remain compliant and competitive in the global market (De Marchi, Di Maria, & Micelli, 2013a; Khan et al., 2019; Navarrete et al., 2020; Arrigo, 2020).

CSOs, local communities, and environmental groups are essential in promoting environmental awareness and advocating for sustainable operations. They exert pressure on suppliers to engage in sustainable practices to maintain their social licenses to operate (De Marchi et al., 2019; Ponte et al., 2023). This grassroots level of engagement ensures that sustainability is not just a regulatory or theoretical concept but a practical and ongoing effort.

Furthermore, industry associations and trade unions contribute by setting operational benchmarks and providing necessary training to comply with these standards. They facilitate collaboration among stakeholders, which enhances the overall efficacy and implementation of ESPs (Ponte et al., 2023; Shen et al., 2017). Simultaneously, academic institutions play a crucial role by advancing the industry's knowledge base through focused research and education on sustainability (Cai & Choi, 2020; De Marchi et al., 2019).

In summary, the integrated efforts of these varied stakeholders crucially advance environmental sustainability among suppliers in the T&A GVCs. By advocating for rigorous standard-setting, enhancing educational outreach, and shaping policy formulation, they collectively drive significant improvements in environmental performance and foster a culture of sustainability within the industry. This collaborative and multifaceted approach underlines the importance of stakeholder synergy in achieving profound environmental sustainability in the T&A sector.

3.2.2 Environmental Sustainability Process Within the T&A GVC

In the preceding section, we explored the various drivers that propel the environmental sustainability process (ESP) and the tactics they employ to encourage or compel suppliers to adopt environmental sustainability measures. Figure 6 delineates the sequence of actions that suppliers undertake upon receiving directives, demands, and strategies for environmental sustainability from diverse stakeholders.

Once suppliers receive environmental sustainability directives or strategies from different stakeholders, they conduct a thorough assessment of their production processes to ascertain their environmental footprint. This involves examining the inputs used, the processes followed, and the outputs produced. Such an analysis not only clarifies the environmental impacts but also assists in tailoring strategies to target specific environmental issues effectively. Subsequently, suppliers plan and pinpoint the precise initiatives necessary to tackle these environmental problems, thereby meeting the expectations of different stakeholders. Following this, they identify any obstacles that may hinder the execution of these initiatives.

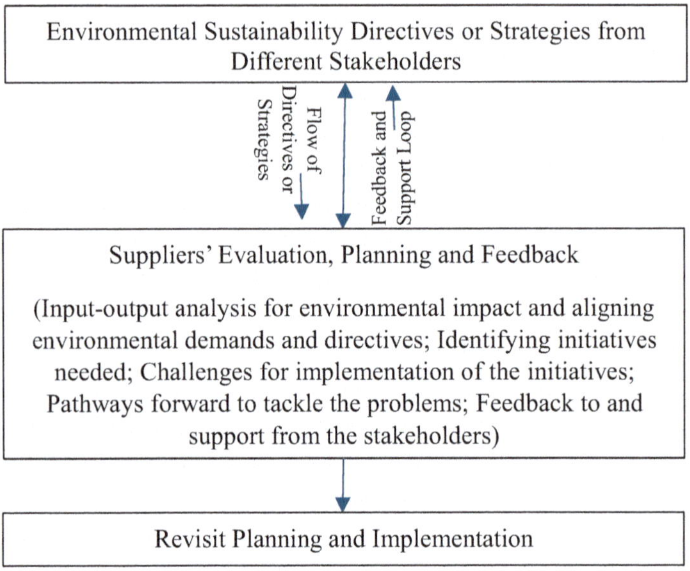

Fig. 6 GVC analytical framework for identifying environmental sustainability process within the T&A GVC (an extension of Fig. 3). (Source: Created by the authors)

The subsequent step involves charting out the pathways to progress. In recognition of the collaborative nature required by the ESP, firms provide feedback regarding the demands, directives, and strategies proposed by various stakeholders and solicit their support. With the requisite backing, suppliers then revisit planning and develop a detailed implementation plan for the identified initiatives. The planning phase is a critical juncture that precedes the execution of sustainability measures, marking a commitment to environmental stewardship in line with stakeholder collaboration. The process concludes with the execution of the plan, which represents the tangible enactment of the ESP strategies.

3.2.2.1 Environmental Sustainability Directives or Strategies from Different Stakeholders

These issues were discussed in the previous section. Please see Sect. 3.2.1 for detailed discussion.

3.2.2.2 Suppliers' Evaluation, Planning, and Feedback

3.2.2.2.1 The Environmental Impact of the T&A GVC

Navigating through the complexities of the T&A GVC unveils the profound environmental implications at each stage, from the initial design to the final consumer's

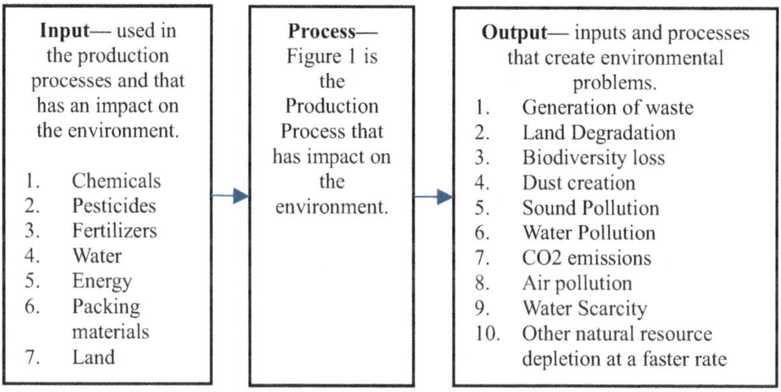

Fig. 7 Input-output analysis for identifying environmental impact within the T&A GVC. (Source: Adapted from Karthik and Gopalakrishnan (2014); Nguyen et al. (2014); De Marchi et al. (2019)

involvement. This exploration delves into the multifaceted environmental impacts inherent within the T&A industry highlighting the challenges. Through a closer examination of each segment of the value chain, as depicted in Fig. 1—Design and Development, Raw Material Production, Textile Manufacturing, Apparel Manufacturing, Distribution and Retail, and the Final Consumer—we gain insights into the industry's ecological footprint.

The textile and apparel manufacturing process extensively uses water, energy, and chemicals, resulting in significant waste and environmental harm due to inefficiency. Capturing the full range of environmental impacts within the T&A GVC is challenging. However, we have developed Fig. 7 by adapting three different sources: Karthik and Gopalakrishnan (2014); Nguyen et al. (2014); De Marchi et al. (2019). The aim of this attempt is to depict both the production process and its environmental impacts succinctly.

The GVC of T&A is complex and multifaceted, with each stage contributing significantly to environmental degradation. Key environmental issues in the textile industry include chemical pollution, excessive water and energy use, air pollution, solid waste, and odor emissions (Islam et al., 2022). Furthermore, Niinimäki et al. (2020) found that the industry causes substantial environmental damage from production to consumption, with annual waste production exceeding 92 million tons and water usage around 79 trillion liters. They recommended critical changes to the T&A business model, such as reducing production speed and implementing sustainable practices throughout the supply chain. Additionally, they suggested that altering consumer behavior to buy fewer clothes and increase the lifespan of garments could also contribute positively.

Here, we delve into the distinct phases of the T&A GVC and examine the specific environmental impacts associated with each.

Design and Development

The journey begins in the design and development phase, where retail brands make crucial decisions that set the environmental course for the entire value chain. Although this stage might not directly produce high levels of pollution, the choice of materials and production techniques has far-reaching implications. For instance, opting for polyester involves intensive water and chemical usage and relies on non-renewable resources, significantly increasing the ecological footprint (Villanueva & Lyu, 2020; De Marchi et al., 2019; Jeppesen & Hansen, 2004). Moreover, the prevalent linear production model of "make, use, dispose" contributes to enormous textile waste, exacerbating pollution and natural resource depletion (Lewis et al., 2017).

Raw Materials Production

Next is the raw materials production stage, which is one of the largest sources of environmental harm (Das, 2021). The cultivation and processing of natural fibers like cotton, as well as synthetic materials such as polyester, demand vast amounts of water, energy, and land. The use of harmful pesticides and chemicals in these processes not only pollutes ecosystems but also poses health risks to human populations (Islam et al., 2022; Felgueiras et al., 2021; Rapsikevičienė et al., 2019).

Textile Manufacturing

During textile manufacturing, which includes processes like fiber production, fabric weaving, dyeing, and finishing, significant environmental challenges arise. This stage is notorious for its heavy chemical use, particularly in dyeing and finishing, leading to substantial water pollution. Additionally, the overall energy consumption and carbon emissions from this phase are remarkably high, contributing to global warming (Niinimäki et al., 2020; Desore & Narula, 2018). As a result, this stage considers as the highest polluter within the T&A GVC (Islam et al., 2022).

Apparel Manufacturing

In apparel manufacturing, which involves cutting, sewing, and finishing, the environmental impacts become increasingly apparent. This phase is characterized by substantial waste generation and energy consumption. The use of non-sustainable materials and inefficient machinery amplifies the industry's carbon footprint, reinforcing the need for more eco-friendly practices. Zhang and Chen (2019) identified that the primary contributors to carbon emissions in garment sewing production were fabrics and accessories. The next major sources were sewing machines, with operators being the third largest contributors to emissions.

Distribution and Retail

The distribution and retail stages also play a significant role in environmental degradation. The logistics of transporting, storing, and selling garments involve large amounts of carbon emissions, particularly due to the reliance on fossil fuels for transportation. The retail phase further contributes to this footprint through energy-intensive store operations and the disposal of unsold or outdated stock, which often ends up in landfills (Gölgeci et al., 2019; De Marchi et al., 2019; Su, 2013; Li et al., 2024).

Final Consumers

Finally, the actions of consumers—how they purchase, use, and dispose of clothing—have direct environmental impacts. Fast fashion trends exacerbate this problem by increasing the amount of waste generated as consumers quickly cycle through clothing. Improper disposal practices and the high resource use in garment maintenance further contribute to the sector's environmental issues (Connell, 2010; Waite, 2014; Mastamet-Mason & More, 2023; Kang et al., 2013).

Conclusion

From design to disposal, each stage of the T&A GVC poses unique environmental challenges. This comprehensive overview highlights the pervasive nature of these impacts, underscoring the critical need for sustainable practices across all facets of the industry. By understanding the specific harms associated with each phase, stakeholders can target interventions more effectively, aiming to mitigate the profound environmental effects inherent in the current textile and apparel production processes. This strategic approach is essential for reducing the overall ecological footprint of the industry and advancing toward a more sustainable future.

3.2.2.2.2 Initiatives for Environmental Sustainability Within the T&A GVC

After the environmental impact has been assessed for each stage of the value chain, the next step is to embark on a journey to identify initiatives aimed at environmental sustainability and improved eco-performance. At each stage—design and development, raw materials production, textile manufacturing, garment manufacturing, distribution, retail, and finally, consumer involvement—initiatives are flourishing. These are aimed at minimizing ecological footprints and fostering a sustainable industry landscape. This narrative unfolds a story of transformation, where innovative practices and strategic collaborations redefine the industry's relationship with the environment. Below, let us delve into these stages and uncover the pivotal initiatives that are shaping and are necessary for a greener future for the T&A GVC.

Design and Development

In the T&A GVC, the design and development stage is critical for embedding environmental sustainability. Leading retail brands are adopting eco-friendly practices early in the product lifecycle to minimize ecological impacts and promote sustainability (Jeppesen & Hansen, 2004). Key initiatives include using sustainable materials like organic cotton, recycled polyester, and Tencel to significantly reduce environmental footprints, alongside implementing energy-efficient and water-saving production techniques (Chan et al., 2016).

Circular design principles are also employed to design products that are durable and recyclable, reducing waste and the need for new resources (Shirvanimoghaddam et al., 2020). Life cycle assessment (LCA) tools are used to evaluate and improve the sustainability of products across all stages of their lifecycle (Dangelico et al., 2013). Collaborative efforts with suppliers and stakeholders ensure these practices are widespread (Gölgeci et al., 2019), while investments in sustainability-focused education prepare future designers to continue these efforts (Sidian et al., 2023).

Raw Materials Production

In the raw materials production stage of the T&A GVC, suppliers, especially in the Global South, are advancing environmental sustainability with innovative strategies. This phase crucially shapes the sustainability standards for the entire chain using both natural and synthetic fibers.

Suppliers are shifting toward sustainable agricultural practices for natural fibers like cotton, wool, and silk, significantly reducing the use of water and harmful pesticides, thus promoting biodiversity and mitigating environmental damage (La Rosa & Grammatikos, 2019). In synthetic fibers, emphasis is growing on recycling to lessen dependency on virgin petrochemicals, with advancements in both chemical and mechanical recycling processes enhancing the circular economy (Alves et al., 2022). In addition, considering the rise of population and consumption patter, we must identify new materials and new process to ensure sustainability within T&A GVC (Felgueiras et al., 2021).

Investment in sustainable raw materials is also increasing, such as the use of agricultural waste to produce viscose rayon fiber, which offers an eco-friendly alternative that decreases the environmental footprint (Nikmatin et al., 2022). Moreover, the introduction of technologies like blockchain for enhancing traceability and transparency in ensuring ethical sourcing and sustainability compliance is contributing to improved sustainability within this stage (Ahmed & MacCarthy, 2021).

Textile Manufacturing

In the textile manufacturing stage of the T&A GVC, suppliers from the Global South are pivotal in adopting sustainable practices to minimize environmental impact. Key strategies include the use of green technologies such as energy-efficient machinery and water-saving dyeing techniques, significantly reducing water, energy, and chemical use (Shen et al., 2017).

Efforts toward zero-waste production are also notable, with innovative practices like using bacterial cellulose to eliminate fabric waste from cutting processes (Chan et al., 2017). Moreover, embracing a circular economy is evident as suppliers recycle used cotton garments into new fibers (Cao et al., 2022) and find beneficial uses for textile sludge in soil enhancement (Abbas et al., 2021).

Apparel Manufacturing

In the apparel manufacturing stage of the T&A GVC, suppliers from the Global South are at the forefront of promoting environmental sustainability. This phase includes activities such as cutting, sewing, packaging, and finishing, which traditionally have a high environmental impact. To combat this, suppliers are implementing innovative green supply chain management strategies to significantly reduce waste, energy use, and carbon emissions (Habib et al., 2022).

Key initiatives include adopting energy-saving technologies like advanced machinery and LED lighting (Hoque et al., 2022), optimizing fabric use, recycling textile waste to support a circular economy (Juanga-Labayen et al., 2022), and switching to renewable energy sources such as solar and wind (Appiadu et al., 2022). Moreover, these suppliers are pursuing sustainability certifications like ISO 14001 and LEED to validate their environmental efforts (De Marchi et al., 2019), and using tools from the Sustainable Apparel Coalition to enhance their operational sustainability (Watson & Wiedemann, 2019).

Distribution and Retail

In the distribution and retail stages of the T&A GVC, firms are actively enhancing environmental sustainability. These companies are focused on reducing their ecological footprints through several strategic initiatives aimed at improving logistics and retail operations.

Efforts include optimizing delivery routes to reduce fuel consumption and emissions, significantly cutting down greenhouse gases (Chargui et al., 2019), and adopting sustainable packaging materials to decrease the use of non-biodegradable resources. Additionally, energy-saving technologies are implemented in logistics centers and retail outlets to reduce energy use, involving upgrades to more efficient lighting and climate control systems.

Moreover, technological advancements support these green initiatives; for example, data analytics optimize inventory management and environmentally friendly packaging solutions minimize waste. The adoption of green logistics principles, such as using electric vehicles and recycling programs, further helps lessen the environmental impact of these activities.

Final Consumers

Final consumers significantly influence environmental sustainability within the T&AGVC. Their decisions impact eco-friendly practices throughout the industry, from acquisition to disposal. As a result, suppliers are taking initiative to educate consumers by innovative communication strategies—such as detailed hangtags and the use of digital platforms—that inform and influence consumer behavior (Brosdahl & Carpenter, 2010; Das, 2021; Li & Leonas, 2021; Mastamet-Mason & More, 2023).

By choosing garments made from organic or recycled materials, consumers drive demand for sustainable products, encouraging brands to adopt greener practices. Thoughtful storage and care extend garment lifespans, reducing consumption and waste. Opting for eco-friendly detergents and colder washes minimizes energy and water use, lowering households' carbon footprints. Furthermore, engaging in responsible disposal methods like donating or recycling old clothes promotes material reuse and supports the circular economy (Connell & Kozar, 2014).

Conclusion

Throughout the T&A GVC, every stage from design and development to final consumer actions is actively enhancing environmental sustainability. Initiatives such as using sustainable materials, implementing green technologies, optimizing logistics, and encouraging responsible consumer behaviors underline a comprehensive industry commitment to reducing ecological footprints. Strategic collaborations and education further embed these practices, setting a foundation for ongoing improvement. This integrated approach not only addresses immediate environmental challenges but also sets a progressive standard for future sustainability in the global marketplace, demonstrating the industry's robust commitment to ecological stewardship.

3.2.2.2.3 Challenges to ESP Implementation Within the T&A GVC

The pursuit of environmental sustainability within the T&A GVC unveils a complex interplay of challenges that span from internal dynamics within the value chain to broader institutional and market-driven barriers. This comprehensive analysis draws insights from empirical studies and strategic discussions to shed light on the multifaceted obstacles to ESP across different sectors, with a particular focus on the textile and apparel industry. The discussion navigates through the intricacies of supplier-lead firm dynamics, financial and technical constraints, the erosion of competitive advantage, internal stakeholder resistance, and the institutional voids that collectively shape the trajectory of ESP in GVCs.

Lack of Support from Lead Firms

In T&A GVCs, while lead firms—mainly large multinational corporations—have the potential to significantly drive ESP, their support is not always consistent or beneficial. These firms often control the sustainability agenda to enhance their green image but might not always wield their influence positively. Research shows that unipolar governance models, where one dominant firm controls the supply chain, can sometimes result in exploitative practices rather than genuine sustainability support. Lead firms may reap the benefits of sustainability efforts from their suppliers without providing adequate compensation, leading to a lack of real commitment to environmental standards down the chain (Poulsen et al., 2016; Jeppesen & Hansen, 2004; Goger, 2013; Khan et al., 2019; Ponte, 2022).

Furthermore, the cascading compliance model that is supposed to ensure adherence to environmental practices across the supply chain often falters. High costs and uncertain financial returns discourage suppliers from adopting necessary environmental practices, with little to no financial support from the lead firms to offset these challenges (Van Assche & Narula, 2023; Raut et al., 2019). This scenario is exacerbated by the use of sustainability as a marketing tool, which can pressure suppliers toward superficial compliance—particularly in developing countries where they may lack adequate financial and technical support (Achabou et al., 2017; Raut et al., 2019).

For true advancement in ESP within T&A GVCs, it is crucial for lead firms to establish transparent and supportive partnerships with their suppliers. This relationship should not only involve fair compensation but also genuine support in implementing sustainable practices to avoid the pitfalls of exploitative governance and ensure that sustainability efforts are authentic and effective.

Suppliers' Financial and Technical Constraints

Suppliers in the Global South within T&A GVC encounter significant obstacles in adopting sustainable practices due to financial and technical constraints. These include scarce resources and structural and operational hurdles, making sustainability progress challenging (Akbar & Ahsan, 2020; Nand et al., 2022; Khattak & Park, 2018; Jeppesen & Hansen, 2004; Khan et al., 2019; Van Assche & Narula, 2023).

Financially, the high costs of ESP and the pressure to adhere to both lead firms' demands and strict environmental standards without sufficient support exacerbate these difficulties (Khan et al., 2019; Achabou et al., 2017). Limited access to advanced technologies necessary for processes like fiber-to-fiber recycling also limits their sustainability capabilities (Singh & Jana, 2023).

Additionally, a lack of technical knowledge and the high cost of acquiring expertise further hinder these suppliers, especially smaller firms or those without strong buyer relationships, from meeting sustainability standards and affording certifications such as LEED (Khattak & Park, 2018; Khattak & Stringer, 2017).

To overcome these challenges and enable suppliers to fully engage in sustainable practices, there is a critical need for increased support from lead firms and other stakeholders in the GVC. This support should include both financial assistance and technical training to build a robust foundation for sustainable development in the industry, fostering an ecosystem where environmental sustainability is not just feasible but thriving.

Trade-Off Between Competitive Advantage and Environmental Sustainability

In T&A GVC, the adoption of ESP often poses a paradox for suppliers, as these initiatives can conflict with maintaining a competitive advantage. This challenge arises from the high costs and strict requirements of ESP, which are compounded by the demanding procurement practices and environmental standards set by lead firms (Tokatli et al., 2008; Tokatli, 2013; Goger, 2013; Khattak et al., 2015; Khattak & Stringer, 2017; Khattak & Park, 2018).

Suppliers also struggle with limited production capabilities and outdated technologies, which further hinder full ESP implementation and exacerbate the trade-off between sustainability investments and market competitiveness (Poulsen et al., 2016). While initial adopters of ESP may experience temporary competitive benefits, these do not always translate into long-term profitability, especially as more players in the market adopt similar strategies (Goger, 2013; Khan et al., 2019).

This complex situation underscores the need for a balanced approach that integrates both economic and environmental objectives without compromising the operational viability of suppliers in the GVC. Such a strategy requires thoughtful consideration of both the immediate financial pressures and the broader environmental responsibilities that suppliers face.

Lack of Support from Internal Stakeholders

Implementing ESP within GVCs faces significant resistance from key internal stakeholders—employees, managers, and investors—who often prioritize short-term gains over long-term sustainability benefits.

Employees are crucial for daily operations affecting environmental performance (Govindarajulu & Daily, 2004; Temminck et al., 2015), yet their resistance to ESP-related changes can impede progress. This resistance typically stems from fears over job security and increased workload, leading to a reluctance to alter established routines (Khan et al., 2019).

Managers, responsible for resource allocation and major decision-making (Nguyen et al., 2014; Desore & Narula, 2018), may also hinder sustainability efforts by focusing on immediate financial outcomes rather than environmental benefits. Influenced by investor pressures, this short-term focus can result in insufficient investments in sustainability, undermining ESP initiatives (Kummritz et al., 2017; Francoeur et al., 2021).

Research indicates that investors might reduce the likelihood of environmental violations in firms and encourage innovation in green technologies through environmental patents. However, other studies show that these investors might also lead to environmental damage through short-term trading behaviors and by funding activities that harm the environment (Yize et al., 2020; Jingbo et al., 2020; Dyck et al., 2019; Dilla et al., 2019). This preference for short-term gains limits financial support for sustainability, complicating its integration within the firm.

In summary, while employees, managers, and investors are vital for the success of ESP within GVCs, their common focus on immediate advantages can create formidable obstacles. Overcoming these challenges requires realigning the interests of all internal stakeholders with the organization's long-term sustainability objectives. This strategic alignment is essential to diminish resistance and promote an organizational culture that supports enduring environmental sustainability.

Institutional Barriers

Implementing ESP within T&A GVCs faces significant institutional barriers from complex governance structures and intense competitive pressures.

Key challenges include the purchasing practices of lead firms that emphasize price competition, leading to cost-cutting measures that often undermine ESP initiatives (Khan et al., 2019). Governance models also affect ESP adoption; unipolar systems with a single dominant firm may push for ESP due to reputational risks, whereas multipolar systems with diverse standards complicate compliance for suppliers (Van Assche & Narula, 2023).

A critical issue is the lack of institutional support and economic incentives, which makes it difficult for suppliers to commit to environmental upgrades. This situation is worsened by inadequate governmental incentives and infrastructure (De Marchi, Di Maria, & Micelli, 2013a; Khattak & Park, 2018), and the pressures to maintain low costs and high quality often deter investments in ESP due to uncertain financial returns (Khan et al., 2019; Milberg & Winkler, 2013).

Furthermore, in the Global South, institutional voids—absence of formal support mechanisms—severely limit effective ESP implementation, especially in critical sectors like textiles and apparel (Khattak, 2013; Khan et al., 2021).

Overall, overcoming these institutional barriers is essential for enabling suppliers in T&A GVCs to effectively implement ESP and achieve sustainable outcomes. This requires a concerted effort to harmonize governance structures, provide adequate support and incentives, and alleviate competitive pressures to foster a more conducive environment for sustainability.

Industry Context

Navigating environmental sustainability within GVCs is particularly complex at the sub-supplier level, where the nuances of sustainability efforts are intensified by unique industry characteristics (Wilhelm et al., 2016; Golgeci et al., 2021; Van Assche & Narula, 2023). This complexity is compounded in industries with specific operational demands that pose distinct sustainability challenges.

For example, industries that extensively use toxic chemicals face significant obstacles in mitigating environmental impacts. Similarly, the textile sector, known for its high water and energy requirements, encounters formidable challenges in adopting sustainable practices. Moreover, the fast fashion industry, characterized by its rapid production cycles, exacerbates these issues by promoting practices that often overlook environmental sustainability in favor of speed and cost-efficiency (Köksal et al., 2017; Ahmed & MacCarthy, 2021).

These industry-specific dynamics highlight the need for customized strategies that address the particular sustainability challenges and environmental footprints of different sectors. Developing and implementing these tailored approaches is crucial for effectively enhancing environmental sustainability across the diverse landscapes of global supply chains.

Lack of Motivation Due to the Sense of Environmental Injustice Among Suppliers from Global South

The relationship between multinational enterprises (MNEs) and their suppliers, particularly those in the Global South, often exhibits a significant disconnect in terms of environmental sustainability practices. This gap primarily stems from how MNEs from advanced economies impose economic and environmental standards on their suppliers without fairly sharing the associated costs, leading to a sense of environmental injustice among suppliers (Clarke & Boersma, 2017; Khattak et al., 2015; Khan et al., 2019; Ponte, 2022).

Suppliers' contributions to environmental governance in GVCs are frequently undervalued, and while MNEs may benefit from green strategies, these often have mixed outcomes for suppliers, with the benefits disproportionately accruing to the MNEs (Jeppesen & Hansen, 2004; Poulsen et al., 2016; De Marchi & Gereffi, 2023; Ponte et al., 2023). This inequity can demotivate suppliers from fully engaging in environmental sustainability initiatives, feeling that they bear the costs without receiving commensurate benefits.

Moreover, the prevailing approach by MNEs often neglects to include suppliers' perspectives in shaping environmental policies, which can lead to further discontent and perceived unfairness (Herrigel et al., 2013; Lechner et al., 2020). Addressing these issues of environmental justice is not only a matter of ethics but also a strategic imperative for MNEs. Recognizing and incorporating supplier viewpoints can improve operational efficiency, promote knowledge sharing, and strengthen overall supply chain relationships, offering economic advantages grounded in principles of bounded rationality and reliability (Kano, 2018; Buckley & Strange, 2015).

To foster a more motivated and equitable GVC environment, it is crucial for MNEs to actively engage with their suppliers, ensuring that sustainability efforts are both fair and beneficial across all levels of the chain. This approach can help align interests, enhance cooperation, and ultimately lead to more effective and just sustainability outcomes.

Lack of Support from Consumers

The effort to implement environmental sustainability within T&A GVC is significantly hindered by a lack of consumer support. This challenge stems from discrepancies between consumer attitudes and their actual purchasing behaviors, often highlighted by the "attitude-behavior gap" where high levels of reported concern about environmental issues do not translate into sustainable purchasing decisions (Bhaduri & Ha-Brookshire, 2011). This gap makes it difficult for companies to invest in sustainable production methods without clear demand for such products.

Additionally, the popularity of the fast fashion model exacerbates sustainability challenges within the T&A GVC. Characterized by rapid production cycles and low costs, fast fashion promotes overconsumption and waste, significantly straining natural resources and increasing pollution (Bick et al., 2018; Niinimäki et al., 2020). This model undermines sustainability efforts as the transient nature of fashion trends encourages consumers to view clothing as disposable.

Another barrier is the low consumer engagement with sustainability labels and certifications, such as the Global Organic Textile Standard (GOTS) or the Fair Wear Foundation. Despite their availability, skepticism about their authenticity and a general lack of awareness limit their effectiveness in guiding consumer choices toward sustainability (Goworek et al., 2012).

Economic considerations also deter consumer support for sustainability; the higher costs associated with sustainably produced apparel can be prohibitive for many shoppers, making price a significant barrier (Joy et al., 2012).

To overcome these obstacles, it is crucial to bridge the attitude-behavior gap, counteract the fast fashion cycle, increase awareness and trust in sustainability certifications, and make sustainable products more economically accessible. Addressing these issues requires a concerted effort from all stakeholders to enhance consumer engagement and demand for sustainable products, which is essential for advancing environmental sustainability in the T&A GVC.

Conclusion

These insights underscore the multifaceted challenges to achieving environmental sustainability within GVCs, highlighting the need for a collaborative approach that includes fair practices, equitable sharing of costs and benefits, active engagement with suppliers, and increased consumer awareness to support the adoption of green practices across the industry.

3.2.2.2.4 Pathways Forward for Ensuring Environmental Sustainability Within the T&A GVC

In the global economy, the textiles and apparel (T&A) global value chain (GVC) plays a crucial role with significant implications for environmental sustainability. While addressing the various environmental harms and outlining the challenges to sustainability, several potential pathways were hinted at. This section will delve deeper into these strategies to provide a comprehensive exploration of how to effectively navigate and overcome these challenges.

Leveraging Lead Firm Influence

For suppliers in the Global South aiming to integrate environmental sustainability within T&A GVCs, leveraging the influence of lead firms can be a key strategy. Lead firms typically possess considerable control and resources, which can be pivotal in supporting suppliers' environmental initiatives. By utilizing their extensive networks, lead firms can provide crucial knowledge, resources, and facilitate connections with certifying bodies, all of which are essential for suppliers to effectively enhance their ecological performance (Khattak et al., 2015; Khan et al., 2019; Van Assche & Narula, 2023). This support not only helps in meeting global sustainability standards but also in aligning suppliers with international market demands, thus enabling a smoother transition into sustainable practices.

Building Internal Capabilities

Developing internal capabilities is critical for suppliers in the Global South to effectively integrate ESP within T&A GVCs (Habib & Bao, 2019; Huo, 2012). This involves enhancing workforce expertise, adopting advanced technologies, and implementing best practices in environmental management to empower firms to adopt and execute sustainable practices robustly.

The transformation seen in the Sri Lankan apparel industry is a testament to the benefits of bolstering internal capabilities, where suppliers have overcome substantial ESP-related challenges by advancing their technical knowledge and organizational processes (Khattak & Stringer, 2016; Khattak et al., 2015). These enhancements have included partnerships with specialist organizations, consultations with international experts, and investments in certifications and audits to build credibility and align with international standards (He et al., 2024).

An example of effective internal capability building is VIYELLATEX in Bangladesh, which has invested in environmental innovation linked directly to employee incentives such as wages and bonuses, fostering a culture of sustainability despite external challenges like lack of governmental support. This initiative has turned their facility into a model "green factory" (Khattak & Park, 2018).

Moreover, embracing international standards and viewing environmental constraints as avenues for competitive advantage are vital. Firms are encouraged to see these standards as opportunities to meet large clients' demands and enhance their market positioning (Baden et al., 2009). This approach not only boosts immediate profitability through corporate social responsibility (CSR) initiatives but also strengthens long-term competitiveness by fostering an entrepreneurial mindset and commitment among all internal stakeholders—managers, employees, and investors (Luken & Stares, 2005; Kadarusman & Nadvi, 2013).

In essence, strengthening internal capabilities is indispensable for suppliers to effectively navigate the complexities of ESP implementation in GVCs. This strategy ensures that suppliers are well equipped to meet the dual challenges of environmental standards and competitive market pressures, ultimately leading to improved sustainability and business success.

Policy Interventions from the Government

Government policy interventions are essential for supporting suppliers in overcoming the challenges associated with integrating ESP within GVCs (Iqbal et al., 2022). Suppliers face numerous obstacles like resource scarcity, limited access to technology, and weak bargaining power, all of which can hinder their ability to implement ESP effectively (Khan et al., 2019).

One critical method of government intervention is the enforcement of environmental regulations, which help set universal standards and encourage firms to adopt eco-friendly practices (Zhou et al., 2022). An example of this is in Hebei, China, where targeted environmental regulations have provided a clear compliance framework for manufacturers, facilitating easier adoption of ESP (Hou et al., 2023).

Governments can also help alleviate financial burdens through subsidies, grants, loans, or tax incentives (Mahmoudi & Rasti-Barzoki, 2018). For instance, the Italian government's financial assistance to the home-furnishing sector has enabled suppliers to manage the costs associated with ESP more effectively (De Marchi, Di Maria, & Micelli, 2013a).

Technical assistance is another vital area of support, which includes training programs, technology transfers, and capacity building. The Sri Lankan government's initiatives to enhance the apparel industry through technical support showcase how such efforts can reduce barriers related to inadequate knowledge and technology (Khattak & Stringer, 2016).

Moreover, government actions can foster better collaboration between suppliers and lead firms. By encouraging joint initiatives and dialogue, governments can help improve suppliers' bargaining power and access to essential resources (Khan et al., 2019; Humphrey & Schmitz, 2002; Ponte et al., 2023).

In conclusion, robust government policy interventions are pivotal in enabling suppliers to navigate the complexities of ESP implementation in GVCs. By providing regulatory guidance and financial and technical support and by promoting cooperative relationships within the industry, governments play a crucial role in boosting environmental performance and competitiveness of suppliers, thus advancing sustainable development goals across the industry.

Role of Final Consumers

To bolster consumer support for environmental sustainability in T&A GVC, a multifaceted strategy is essential. Initiatives should focus on educating consumers about the environmental implications of their purchases and the importance of sustainability labels, potentially bridging the prevalent attitude-behavior gap (Henninger et al., 2016). Enhancing transparency within the T&A GVC can also improve trust in sustainability claims, making consumers more likely to support eco-friendly products.

The rise of digital platforms and social media offers significant opportunities for engaging with consumers and promoting sustainability issues more effectively. However, the persistent demand for fast fashion—characterized by low costs and high environmental costs—underlines the need for raising awareness about the ecological damages of such consumption patterns and encouraging willingness to pay more for sustainably produced goods (Li et al., 2024).

Making Environmental Sustainability a "Business Case"

For suppliers, particularly in the Global South, it is crucial to align environmental sustainability with economic viability to make it a compelling business case (Uddin et al., 2023). Emphasizing sustainable procurement that balances environmental and economic considerations can offer a competitive edge and enhance overall performance (Laosirihongthong et al., 2019; Khattak et al., 2015).

Despite potential cost savings and favorable conditions for sustainability in regions like Sri Lanka, the business rationale for investing in ESP often remains unclear due to the lack of guaranteed returns on investment (Goger, 2013). This uncertainty may render sustainability a less appealing investment choice. Additionally, research indicates that any cost savings achieved through eco-innovation by suppliers are often appropriated by lead firms (Jensen & Whitfield, 2022), further complicating the financial incentives for suppliers to invest in these practices.

The focus of many MNEs on purely environmental metrics—such as reducing waste and carbon emissions—can sometimes neglect the financial and relational aspects of business operations, making sustainability efforts seem exploitative to suppliers (Clarke & Boersma, 2017). To truly foster environmental sustainability, it is essential that MNEs integrate financial performance and quality of inter-firm relationships into their sustainability strategies (Buckley & Strange, 2015; Kano, 2018; Krishnan et al., 2023; Van Assche & Narula, 2023).

This integrated approach should ensure that environmental initiatives are not only about compliance but also about fostering genuine partnerships, aligning financial and environmental goals, and building a robust business case for sustainability across the GVC.

Customization of Sustainability Practices

Environmental sustainability in T&A GVCs requires a nuanced and tailored approach. It is crucial for MNEs, governments, and all involved stakeholders to understand that sustainability practices must be specifically adapted to suit the distinct phases of the GVC (De Marchi, Di Maria, & Ponte, 2013b; Khattak et al., 2015; De Marchi & Gereffi, 2023). Each segment, from textile production to garment manufacturing, has its own unique challenges and priorities that need to be addressed (De Marchi & Gereffi, 2023; Krishnan et al., 2023; De Marchi et al., 2019).

Acknowledging these variances is key to customizing sustainability efforts effectively. By aligning sustainability measures with the specific environmental and operational realities of each segment within the GVC, stakeholders can ensure that these initiatives are not only more effective but also directly relevant. This strategic customization helps to unify and significantly advance environmental sustainability efforts across the entire value chain.

Embracing Circular Economy in Fashion

Adopting a circular economy (CE) model in the fashion industry represents a significant paradigm shift aimed at extending the lifecycle of materials through recycling, reuse, and efficient resource management (Thinakaran et al., 2023; de Lima et al., 2022; Suarez-Visbal et al., 2024). Moving away from the traditional "take-make-dispose" model, the CE approach promotes a system where materials are perpetually recycled, minimizing waste and enhancing resource efficiency (Moreau et al., 2017; Charnley et al., 2024).

Despite growing consumer interest in sustainable clothing, significant hurdles remain in fully implementing circular practices. One major challenge is the limited availability of textile-to-textile (T-T) recycling facilities, which currently account for only a small fraction of global textile production (Charnley et al., 2024; de Hugo et al., 2022; Marques et al., 2020). To truly foster a sustainable fashion ecosystem, it is essential to expand the T-T recycling sector.

The path to a robust circular economy in fashion requires a comprehensive strategy that includes promoting social entrepreneurship, reducing waste, encouraging upcycling, and maintaining a firm commitment to CE principles (Atstja et al., 2021). Collaboration among various stakeholders is necessary to drive substantial change and achieve a just and sustainable circular economy within the T&A GVC. Moreover, leveraging digital innovations can empower smaller, eco-conscious brands to make impactful contributions to circularity (Ki et al., 2021).

This approach must also emphasize the shared responsibilities of both consumers and corporations in nurturing a circular economy, highlighting the importance of collective action and ethical commitment in overcoming environmental challenges within the fashion industry.

Collaboration and Partnerships

Environmental collaboration—encompassing joint goal setting, shared planning, and partnership efforts—plays a critical role in enhancing organizational environmental effectiveness across GVCs (Vachon & Klassen, 2008; Gölgeci et al., 2019; Van Assche & Narula, 2023; Ponte et al., 2023; Krishnan et al., 2023). This collaborative approach is vital for interorganizational relations and is considered pivotal for economic and social progress (Achabou et al., 2017; Harari, 2014).

Despite potential conflicts and the complexity of collaboration, the benefits typically outweigh the challenges. These partnerships often involve environmental initiatives, capacity building, and the exchange of best practices between suppliers and lead firms, which have significantly advanced sustainability in various industries worldwide—from apparel in Sri Lanka to home furnishings in Italy (De Marchi, Di Maria, & Ponte, 2013b; Khattak & Stringer, 2016; Goger, 2013).

Financial and technical support through these partnerships, as demonstrated in the South African wine industry (Ponte, 2021), and advocacy for stringent environmental regulations also form a crucial part of collaborative efforts. Moreover, arrangements like cost-sharing agreements within the T&A GVC have been shown to improve environmental efficiency and profitability for both manufacturers and retailers (Li et al., 2024).

Collaborative networks also extend beyond traditional stakeholders, incorporating external aid and institutional support, as seen in the Tunisian olive oil sector and the Indian leather industry. These efforts help achieve compliance and enhance sustainability (Achabou et al., 2017; Tewari, 2006). Furthermore, initiatives like the Sustainable Apparel Forum (SAF) in Bangladesh's garment industry highlight the collective commitment to integrating economic, social, and environmental objectives (Sustainable Apparel Forum, 2023).

However, there is often an overemphasis on the role of lead firms, which can overshadow the essential contributions of smaller, multi-layered suppliers within the GVC. Recognizing and valuing these suppliers' roles is crucial for a balanced approach to sustainability and for achieving widespread environmental improvements across the industry. This holistic perspective on collaboration underscores the need for equitable and inclusive partnership strategies that engage all levels of the supply chain in meaningful sustainability efforts (Chen & Burns, 2006).

3.2.2.2.5 Feedback to and Support from Stakeholders

In the textile and apparel global value chain, suppliers play a crucial role in implementing environmental directives and strategies received from various stakeholders. These directives often focus on sustainable practices, social responsibility, and environmental management within the supply chain. Suppliers receive support from stakeholders such as lead firms, regulatory bodies, and consumers once feedback on these directives is communicated.

Lead firms in the textile and apparel industry are increasingly emphasizing sustainability issues and are working closely with suppliers to ensure compliance with environmental standards (Shen et al., 2017; Krishnan et al., 2023). These lead firms provide guidance, resources, and incentives to suppliers to adopt sustainable practices and reduce their environmental impact. Regulatory bodies also play a significant role by setting standards and regulations that suppliers must adhere to in terms of environmental management and social responsibility (Cai & Choi, 2020). Compliance with these regulations is essential for suppliers to maintain their operations and access global markets.

Moreover, consumers are becoming more environmentally conscious and are driving the demand for sustainable products in the textile and apparel industry (Curwen et al., 2013). Suppliers who align with consumer preferences for eco-friendly products receive support in the form of increased business opportunities and market access. Consumer demand acts as a catalyst for suppliers to improve their environmental performance and implement sustainable practices throughout the value chain.

Overall, suppliers in the textile and apparel global value chain receive support from lead firms, regulatory bodies, and consumers once feedback on environmental directives is communicated. This support comes in the form of guidance, resources, incentives, market opportunities, and regulatory compliance measures, all aimed at promoting sustainability and responsible environmental practices within the industry.

3.2.2.3 Revisit Planning and Implementation of ESP Strategies

Following the assessment and preparatory stages, the "Revisit Planning and Implementation" phase is the crucial step where suppliers take the feedback and analysis outcomes to refine and enhance their ESP. This involves revisiting initial plans with a keen eye for detail and adaptability, ensuring that strategies are not only sound in theory but also feasible in practice.

Suppliers must address the feasibility of the proposed initiatives, considering the identified barriers to implementation. It is a phase of recalibration where plans may be adjusted to reflect the realities of available resources, technological constraints, and the operational environment. It is also a stage for innovation, where alternative methods and novel solutions are considered to overcome the challenges previously identified.

This phase is dynamic and iterative, requiring ongoing dialogue with stakeholders to ensure that the evolving plans remain aligned with the overall sustainability goals and stakeholder expectations. Suppliers may need to engage in further capacity-building, secure additional resources, or even seek new collaborations to address gaps in expertise or technology.

The implementation plan developed in this phase must be comprehensive, including timelines, milestones, resource allocation, and defined responsibilities. It sets the stage for action and holds all parties accountable. This stage culminates in the initiation of the execution phase, where the strategies are put into operation. The success of this phase is contingent on the commitment to continuous improvement and willingness to adapt strategies in response to both expected and unforeseen challenges during execution.

It is through this detailed and reflective approach to planning and implementation that suppliers can effectively turn the vision of environmental sustainability into a practical and actionable reality within their operations.

3.2.3 Environmental Performance Within the T&A GVC

ESP within the T&A GVC is instrumental in promoting industry greening and enhancing the environmental footprint of products. These practices are acknowledged for their capacity to yield a wide array of environmental outcomes, which are critical for the sustainable development of GVCs.

The primary challenges that ESP aim to address in T&A GVC encompass a broad range of issues, from the extensive use of harmful chemicals to the enormous expenditure of water and energy. Additionally, these practices focus on reducing the substantial generation of solid and gaseous wastes and land degradation, curtailing the vast fuel consumption required for transporting goods to remote textile facilities, and minimizing the use of non-biodegradable packaging materials. These problems extend all the way from resource usage and waste production in manufacturing to the carbon and greenhouse gas emissions involved in reaching the consumer (Yao & Geng, 2015; Choudhury, 2014; Costa et al., 2020; Islam et al., 2022; Niinimäki et al., 2020).

The efficacy of ESPs leads to varying positive environmental outcomes. Direct benefits include observable improvements in resource utilization and pollution control. In contrast, indirect outcomes, such as mitigating land degradation and fostering biodiversity, emerge over longer periods, illustrating the layered complexity of implementing and evaluating ESPs within GVCs (Goger, 2013; Golini et al., 2018).

Positive impacts are observed through cleaner production technologies and product improvement initiatives, leading to reduced emissions and a decrease in waste. Such advancements not only improve the environmental performance of individual suppliers but can also enhance the overall environmental footprint of the entire value chain (Khattak et al., 2015; Khattak & Park, 2018). However, the anticipated environmental benefits of ESPs are not always realized uniformly across the board (Dauvergne & Neville, 2009; Bolwig et al., 2010; Lund-Thomsen, 2004; Barrientos, 2019; Krishnan et al., 2023; Golini et al., 2018).

Nevertheless, these initiatives can also introduce complexities. The adoption of environmentally friendly, yet costlier, technologies by some suppliers may pressure others within the value chain to compromise on environmental standards to manage costs, potentially worsening the overall environmental performance (Goger, 2013; De Marchi & Gereffi, 2023). This scenario aligns with the "pollution havens" phenomenon, where strict environmental regulations in certain jurisdictions inadvertently push pollutive activities to less regulated areas (Eskeland & Harrison, 2003).

Furthermore, the impact of these initiatives can vary significantly across the GVC, being beneficial in one area while detrimental in another. This duality highlights the nuanced relationship between environmental sustainability processes and overall environmental performance within the T&A GVC, influenced by governance structures, initiative specifics, and the complexities of supplier markets (De Marchi & Gereffi, 2023).

Figure 8 proposes a comprehensive approach to assess the environmental performance of ESPs, underlining the importance of ongoing evaluation and enhancement.

Figure 8 represents a systematic approach to implementing ESP within the T&A GVC. It outlines a series of steps starting from the implementation of ESP initiatives to the final reporting and communication of results.

The process begins with the "Implementation of ESP," marking the initiation of previously outlined sustainability plans. This is followed by "Measuring results both

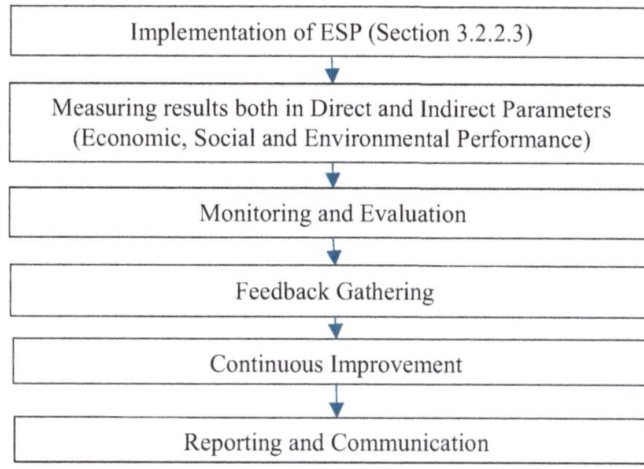

Fig. 8 GVC analytical framework for evaluating environmental performance within the T&A GVC (an extension of Fig. 3). (Source: Created by the authors)

in Direct and Indirect Parameters," a critical step that evaluates the economic, social, and environmental impacts of the practices implemented. This assessment is essential for determining the effectiveness of the ESPs and for gaining a comprehensive understanding of their outcomes, considering both immediate (direct) and broader (indirect) effects. However, existing research often fails to address the environmental impact across all planetary boundaries, which is a significant oversight that should be rectified (De Marchi & Gereffi, 2023; Krishnan et al., 2023; De Marchi et al., 2019).

Subsequent to the measurement phase is "Monitoring and Evaluation," a critical ongoing process that ensures that the ESPs are not only implemented but also maintained and improved upon. This step involves regular checks to assess compliance with planned actions, to evaluate the effectiveness of these actions, and to identify areas for enhancement.

Following the evaluation is "Feedback Gathering." This is where insights and critiques are collected from all relevant stakeholders. This feedback is essential for understanding the efficacy of ESPs from multiple perspectives and for identifying potential areas of improvement.

The next phase is "Continuous Improvement," which is about taking the insights gained from feedback and evaluation to make iterative improvements to the ESPs. It reflects a commitment to ongoing development rather than assuming that initial implementation is the last step.

Finally, "Reporting and Communication" concludes the cycle, where the outcomes, challenges, and knowledge gained from the ESP implementation are documented and communicated to stakeholders. This step is vital for transparency, for learning, and for garnering support for future sustainability initiatives.

4 The Role of Environmental Sustainability Within T&A GVC in Achieving Key SDGs

Environmental sustainability within T&A GVC is increasingly recognized as a key contributor to achieving several United Nations Sustainable Development Goals (UN SDGs). By embedding sustainable practices throughout its operations, the T&A sector can address specific environmental harms and yield a wide array of beneficial outcomes, both direct and indirect, from improved resource efficiency and waste reduction to the broader implications on biodiversity conservation and ecosystem services.

In terms of water conservation, the industry's efforts align with SDG 6, which is centered on clean water and sanitation (Sinha et al., 2021; Akhtar & Suki, 2022). The reduction of wastewater and the optimization of energy efficiency in production processes lay the groundwork for sustainable development within the industry. Recycling methods for textile waste contribute to this goal by ensuring the conservation of water resources (Baloyi, 2023).

Focusing on energy issues, the commitment of the textiles and apparel (T&A) industry to sustainable practices contributes to Sustainable Development Goal 7, which promotes affordable and clean energy (Arrigo, 2020; Sinha et al., 2021). The adoption of sustainable sourcing and efficient supply chain management practices helps to enhance energy efficiency in production processes, aligning with this SDG. Furthermore, key measures including the implementation of energy-conserving technologies such as advanced machinery and LED lighting (Hoque et al., 2022), as well as the shift toward renewable energy sources like solar and wind (Appiadu et al., 2022), support the achievement of SDG 7.

Regarding production, the industry plays a significant role in advancing SDG 12, focused on responsible consumption and production. Incorporating sustainable strategies in product design and resource management emphasizes the industry's role in fostering responsible practices (Shen et al., 2017; Akhtar & Suki, 2022; Vićentijević et al., 2022). Furthermore, efforts to minimize the ecological footprint of production processes, such as utilizing water-saving technologies and embracing the digital economy, directly contribute to responsible consumption and production patterns (Xiang-mei et al., 2022). In addition, traceability and transparency in the textile and clothing sector help establish trust among stakeholders and support the attainment of Sustainable Development Goal 12 (Papú Carrone, 2020).

Climate action, as outlined in SDG 13, is another critical area where the T&A industry's environmental sustainability efforts are crucial. Green practices that reduce material waste and integrate eco-friendly technologies are at the forefront of efforts to mitigate climate change (de Nariê Rinke Dias et al., 2022; Sinha et al., 2021; Akhtar & Suki, 2022). Circular fashion models that promote textile reuse and recycling offer strategies to minimize the industry's environmental footprint and enhance climate action efforts (Moreau et al., 2017; Charnley et al., 2024).

Finally, SDG 15, which focuses on life on land, benefits from the T&A industry's commitment to environmental sustainability. The sector's dedication to reducing material waste and implementing eco-friendly innovations is pivotal in safeguarding land ecosystems and biodiversity (Fei et al., 2021). The exploration of bio-based processes and sustainable education initiatives represents the industry's endeavor to preserve terrestrial habitats (Reis et al., 2020).

Through strategic efforts that span from conserving water resources to reducing carbon emissions and promoting renewable energy, the T&A sector not only improves its own environmental footprint but also makes significant contributions toward a sustainable future aligned with the SDGs. These multifaceted approaches, bolstered by technological advancements and innovative production methods, underscore the integral role of the T&A industry in promoting environmental sustainability within the broader context of global value chains and sustainable development.

5 Conclusion

This chapter presents a pivotal contribution to the understanding and advancement of environmental sustainability within the textile and apparel global value chain (T&A GVC), especially through the lens of Global South's participation. By unveiling a novel analytical GVC framework complemented by three key extensions, this work enriches the discourse on environmental sustainability, integrating the diverse perspectives of international business (IB), global value chains (GVC), and environmental justice (EJ). This comprehensive approach not only elevates academic dialogue but also offers tangible insights for practitioners, policymakers, and other stakeholders engaged in the T&A industry globally.

Central to the chapter's achievements is the innovative application of the GVC framework's six dimensions—input-output structures, geographic and industrial reach, governance mechanisms, paths for upgrading, industry stakeholders, and institutional context. This detailed exploration facilitates a nuanced understanding of environmental sustainability's operationalization within the T&A GVC, spotlighting the critical roles of governance and supplier upgrading processes. Through this lens, the narrative underscores the paramount importance of collaborative endeavors among a wide array of stakeholders, from leading firms to end consumers, in driving forward the agenda for environmental sustainability within the GVC. Additionally, our novel analytical GVC framework questions the conventional emphasis of prior studies, which generally analyze the role of lead firms through GVC governance dimensions and the involvement of suppliers via upgrading pathways within GVCs. We illustrate the critical need to examine all dimensions simultaneously when investigating environmental sustainability for suppliers from the Global South. Our results indicate that adopting an integrated approach is crucial for a genuine comprehension and improvement of the collective benefits associated with GVC participation, thus promoting holistic economic, social, and environmental progress.

This crafted framework elucidates the intricate dynamics among stakeholders, emphasizing the necessity of collective action to surmount the challenges and leverage the opportunities for environmental sustainability. It meticulously outlines the strategies through which sustainability is governed and navigated, offering insights into the complex interplay between diverse actors, particularly highlighting the contributions and impacts on suppliers in the Global South.

In doing so, the chapter addresses the crucial need for continuous assessment and enhancement of environmental sustainability processes (ESPs) to achieve desired environmental performance (EP) outcomes. It recognizes the evident environmental benefits of such practices, yet also acknowledges the hurdles in their uniform application across the value chain. This acknowledgment calls for a steadfast commitment to refining ESPs, ensuring they are effectively tailored to meet the dynamic and varied nature of environmental challenges within the T&A GVC.

By seamlessly integrating theoretical frameworks with practical implications, this chapter lays a robust foundation for future research and initiatives aimed at aligning industry practices with environmental sustainability and the broader objectives of the UN SDGs. It marks a significant stride in the discourse on fostering an environmentally sustainable and equitable T&A GVC, especially considering the evolving role of the Global South in this vital global industry. The insights and analytical tools provided here not only contribute to academic knowledge but also guide the industry toward more sustainable, equitable, and efficient practices, shaping a path toward a greener and more inclusive future. We expect that the analytical framework discussed in this chapter will guide future researchers in understanding how to examine environmental sustainability within any global value chains.

6 Future Research Areas

As the global textile and apparel industry confronts the urgent necessity for environmental sustainability, especially within the Global South, a nuanced, multifaceted approach becomes indispensable. This necessity uncovers substantial gaps in existing research, delineating several areas ripe for further scholarly exploration.

First, future research should endeavor to take a deeper examination of the roles and impacts of multi-layered suppliers within the GVC, particularly those in the Global South. Investigating their contributions to environmental sustainability and GVC governance could reveal novel avenues for promoting sustainable practices across the industry.

Second, extending research to include the development and application of planetary boundaries in environmental sustainability assessments within the textile and apparel industry is vital. Establishing industry-specific metrics for environmental initiatives could lead to more customized and effective sustainability strategies.

Third, the application and implications of environmental justice (EJ) within the textile and apparel industry warrant thorough investigation. Focusing on distributive, procedural, recognition, and other types of justice for suppliers from the Global South presents a significant area for future research.

Fourth, there is a need for further exploration of effective governance models that harmonize environmental sustainability with operational efficiency. This includes scrutinizing how MNEs and emerging country MNEs engage with and support their suppliers in adopting sustainable practices.

Fifth, the dynamic between consumer behavior and sustainable practices within the textile and apparel value chain offers a fertile area for investigation. Investigations could focus on how consumer demand for sustainable products affects supplier practices and sets industry benchmarks. Additionally, exploring strategies that business firms might use to raise awareness among consumers and encourage them to pay a premium for environmentally friendly products would also be a valuable avenue for study.

Finally, exploring the role of technological innovations in promoting environmental sustainability in the textile and apparel industry can provide insights into future strategies for waste reduction, energy efficiency, and circular economy practices.

By tackling these critical areas of research, scholars and industry stakeholders can collaboratively chart pathways toward a more sustainable, equitable, and resilient textile and apparel industry, with a particular emphasis on addressing the challenges and opportunities in the Global South.

References

Abbas, M. M., Nazir, A., Nasr, A. S., & Darweish, M. A. (2021). Utilization of industrial wastes for improving geotechnical properties of collapsible soil. *Journal of Engineering Research*. https://doi.org/10.21608/erjeng.2021.53448.1001

Achabou, M. A., Dekhili, S., & Hamdoun, M. (2017). Environmental upgrading of developing country firms in global value chains. *Business Strategy and the Environment, 26*(2), 224–238. https://doi.org/10.1002/bse.1911

Ahmed, W. A., & MacCarthy, B. L. (2021). Blockchain-enabled supply chain traceability in the textile and apparel supply chain: A case study of the fiber producer, Lenzing. *Sustainability*. https://doi.org/10.3390/su131910496

Akbar, S., & Ahsan, K. (2020). Investigation of the challenges of implementing social sustainability initiatives: A case study of the apparel industry. *Social Responsibility Journal*. https://doi.org/10.1108/srj-09-2019-0291

Akhtar, M. F., & Suki, N. M. (2022). 'Green consumer behaviour: Integration of theory of planned behaviour and technology acceptance model. In *Research Anthology on measuring and achieving sustainable development goals* (pp. 863–871). IGI Global.

Alves, L., Ferreira Cruz, E., Lopes, S. I., Faria, P. M., & Rosado da Cruz, A. M. (2022). Towards circular economy in the textiles and clothing value chain through blockchain technology and IoT: A review. *Waste Management & Research, 40*(1), 3–23.

Appiadu, D., Senayah, W. K., & Biney-Aidoo, V. (2022). Ghana's textile and apparel value chain: Analysis of enabling business environment. *African Journal of Applied Research*. https://doi.org/10.26437/ajar.31.10.2022.07

Arrigo, E. (2020). Global sourcing in fast fashion retailers: Sourcing locations and sustainability considerations. *Sustainability*. https://doi.org/10.3390/su12020508

Atstja, D., Cudečka-Puriņa, N., Vesere, R., Abele, L., & Spivakovskyy, S. (2021). Challenges of textile industry in the framework of Circular Economy: case from Latvia. In *International Conference on Sustainable, Circular Management and Environmental Engineering (ISCMEE 2021)*. EDP Sciences.

Baden, D. A., Harwood, I. A., & Woodward, D. G. (2009). The effect of buyer pressure on suppliers in SMEs to demonstrate CSR practices: An added incentive or counter productive? *European Management Journal, 27*(6), 429–441.

Baloyi, R. B. (2023). Recent advances in recycling technologies for waste textile fabrics: A review. *Textile Research Journal*. https://doi.org/10.1177/00405175231210239

Barrientos, S. (2019). *Gender and work in global value chains: Capturing the gains?* Cambridge University Press.

Barrientos, S., Gereffi, G., & Rossi, A. (2011). Economic and social upgrading in global production networks: A new paradigm for a changing world. *International Labour Review, 150*(3–4), 319–340. https://doi.org/10.1111/j.1564-913X.2011.00119.x

BGMEA. (2020). *Bangladesh: Home to world's highest number of green garment factories*. . Online. Available at: https://www.bgmea.com.bd/page/Sustainability_Environment (Accessed: 10th April 2024).

Bhaduri, G., & Ha-Brookshire, J. E. (2011). Do transparent business practices pay? Exploration of transparency and consumer purchase intention. *Clothing and Textiles Research Journal, 29*(2), 135–149.

Bick, R., Halsey, E., & Ekenga, C. C. (2018). The global environmental injustice of fast fashion. *Environmental Health, 17*, 1–4.

Bolwig, S., Ponte, S., Du Toit, A., Riisgaard, L., & Halberg, N. (2010). Integrating poverty and environmental concerns into value-chain analysis: A conceptual framework. *Development Policy Review, 28*(2), 173–194. https://doi.org/10.1111/j.1467-7679.2010.00480.x

Boström, M., & Micheletti, M. (2016). Introducing the sustainability challenge of textiles and clothing. *Journal of Consumer Policy, 39*(4), 367–375.

Bressán, J. M. (2022). Capitalist crisis and the geography of the clothing industry. *Human Geography*. https://doi.org/10.1177/19427786221120933

Brosdahl, D. J., & Carpenter, J. M. (2010). Consumer knowledge of the environmental impacts of textile and apparel production, concern for the environment, and environmentally friendly consumption behavior. *Journal of Textile and Apparel, Technology and Management, 6*(4).

Buckley, P. J., & Strange, R. (2015). The governance of the global factory: Location and control of world economic activity. *Academy of Management Perspectives, 29*(2), 237–249.

Cai, Y.-J., & Choi, T. M. (2020). A United Nations' sustainable development goals perspective for sustainable textile and apparel supply chain management. *Transportation Research Part E Logistics and Transportation Review*. https://doi.org/10.1016/j.tre.2020.102010

Cao, H., Cobb, K., Yatvitskiy, M., Wolfe, M. L., & Shen, H. (2022). Textile and product development from end-of-use cotton apparel: A study to reclaim value from waste. *Sustainability*. https://doi.org/10.3390/su14148553

Chan, H. K., Yee, R. W., Dai, J., & Lim, M. K. (2016). The moderating effect of environmental dynamism on green product innovation and performance. *International Journal of Production Economics*. https://doi.org/10.1016/j.ijpe.2015.12.006

Chan, C. K., Shin, J. S., & Jiang, S.-X. K. (2017). Development of tailor-shaped bacterial cellulose textile cultivation techniques for zero-waste design. *Clothing and Textiles Research Journal*. https://doi.org/10.1177/0887302x17737177

Chargui, T., Bekrar, A., Reghioui, M., & Trentesaux, D. (2019). Multi-objective sustainable truck scheduling in a rail–road physical internet cross-docking hub considering energy consumption. *Sustainability*. https://doi.org/10.3390/su11113127

Charnley, F., Cherrington, R., Mueller, F., Jain, A., Nelson, C., Wendland, S., & Ventosa, S. (2024). Retaining product value in post-consumer textiles: How to scale a closed-loop system. *Resources, Conservation and Recycling, 205*, 107542.

Chen, H.-L., & Burns, L. D. (2006). Environmental analysis of textile products. *Clothing and Textiles Research Journal, 24*(3), 248–261.

Cheung, M.-S., Myers, M. B., & Mentzer, J. T. (2010). Does relationship learning Lead to relationship value? A cross-national supply chain investigation. *Journal of Operations Management*. https://doi.org/10.1016/j.jom.2010.01.003

Choudhury, A. K. R. (2014). Environmental impacts of the textile industry and its assessment through life cycle assessment. In S. S. Muthu (Ed.), *Roadmap to sustainable textiles and clothing: Environmental and social aspects of textiles and clothing supply chain* (pp. 1–39). Springer Singapore.

Clarke, T., & Boersma, M. (2017). The governance of global value chains: Unresolved human rights, environmental and ethical dilemmas in the apple supply chain. *Journal of Business Ethics, 143*, 111–131.

Connell, K. Y. H. (2010). Internal and external barriers to eco-conscious apparel acquisition. *International Journal of Consumer Studies*. https://doi.org/10.1111/j.1470-6431.2010.00865.x

Connell, K. Y. H., & Kozar, J. M. (2014). Environmentally sustainable clothing consumption: Knowledge, attitudes, and behavior. In S. S. Muthu (Ed.), *Roadmap to sustainable textiles and clothing: Environmental and Social Aspects of Textiles and Clothing Supply Chain* (pp. 41–61). Springer Singapore.

Costa, C., Azoia, N. G., Silva, C. J., & Marques, E. F. (2020). Textile industry in a changing world. *U Porto Journal of Engineering*. https://doi.org/10.24840/2183-6493_006.002_0008

Curwen, L. G., Park, J., & Sarkar, A. K. (2013). Challenges and solutions of sustainable apparel product development: A case study of Eileen Fisher. *Clothing and Textiles Research Journal, 31*(1), 32–47.

Dangelico, R. M., Pontrandolfo, P., & Pujari, D. (2013). Developing sustainable new products in the textile and upholstered furniture industries: Role of external integrative capabilities. *Journal of Product Innovation Management*. https://doi.org/10.1111/jpim.12013

Das, D. (2021). An empirical approach to investigate the possibility of consumer driven sustainability transformation of Indian textile and apparel industries. *Turkish Journal of Computer and Mathematics Education (TURCOMAT), 12*(7), 91–98.

Dauvergne, P., & Neville, K. J. (2009). The changing north–south and south–south political economy of biofuels. *Third World Quarterly, 30*(6), 1087–1102.

de Hugo, A., de Nadae, J., & de Lima, R. S. (2022). Consumer perceptions and actions related to circular fashion items: Perspectives of young Brazilians on circular economy. *Waste Management & Research the Journal for a Sustainable Circular Economy*. https://doi.org/10.1177/0734242x221122571

de Lima, F. A., Seuring, S., & Sauer, P. C. (2022). A systematic literature review exploring uncertainty management and sustainability outcomes in circular supply chains. *International Journal of Production Research, 60*(19), 6013–6046.

De Marchi, V., & Alford, M. (2022). State policies and upgrading in global value chains: A systematic literature review. *Journal of International Business Policy, 5*(1), 88–111. https://doi.org/10.1057/s42214-021-00107-8

De Marchi, V., & Di Maria, E. (2019). Environmental upgrading and suppliers' agency in the Leather Global Value Chain. *Sustainability, 11*(23), 17. https://doi.org/10.3390/su11236530

De Marchi, V., & Gereffi, G. (2023). Using the global value chain framework to analyse and tackle global environmental crises. *Journal of Industrial and Business Economics, 50*(1), 149–159. https://doi.org/10.1007/s40812-022-00253-x

De Marchi, V., Di Maria, E., & Micelli, S. (2013a). Environmental strategies, upgrading and competitive advantage in global value chains. *Business Strategy and the Environment, 22*(1), 62–72. https://doi.org/10.1002/bse.1738

De Marchi, V., Di Maria, E., & Ponte, S. (2013b). The greening of global value chains: Insights from the furniture industry. *Competition & Change, 17*(4), 299–318.

De Marchi, V., Di Maria, E., & Ponte, S. (2014). Multinational firms and the management of global networks: Insights from global value chain studies. In *Orchestration of the global network organization* (pp. 463–486). Emerald Group Publishing Limited.

De Marchi, V., Di Maria, E., Krishnan, A., & Ponte, S. (2019). Environmental upgrading in global value chains. In S. Ponte, G. Gereffi, & G. Raj-Reichert (Eds.), *Handbook on global value chains* (pp. 310–323). Edward Elgar Publishing Limited.

De Marchi, V., Di Maria, E., Golini, R., & Perri, A. (2020). Nurturing international business research through global value chains literature: A review and discussion of future research opportunities. *International Business Review, 29*(5), 101708.

de Narié Rinke Dias, S., Souza, A., Chagas, M. F., Hernandes, T. A. D., & Cavalett, O. (2022). Addressing the contributions of electricity from biomass in Brazil in the context of the sustainable development goals using life cycle assessment methods. *Journal of Industrial Ecology*. https://doi.org/10.1111/jiec.13242

Desore, A., & Narula, S. A. (2018). An overview on corporate response towards sustainability issues in textile industry. *Environment, Development and Sustainability, 20*, 1439–1459.

Diabat, A., & Govindan, K. (2011). An analysis of the drivers affecting the implementation of green supply chain management. *Resources, Conservation and Recycling, 55*(6), 659–667.

Diabat, A., Kannan, D., & Mathiyazhagan, K. (2014). Analysis of enablers for implementation of sustainable supply chain management–A textile case. *Journal of Cleaner Production, 83*, 391–403.

Dilla, W., Janvrin, D., Perkins, J., & Raschke, R. (2019). Do environmental responsibility views influence investors' use of environmental performance and assurance information? *Sustainability Accounting, Management and Policy Journal, 10*(3), 476–497.

Divrik, B., & Baykal, E. (2022). A qualitative study about internationalization of Turkish textile & clothing industries. *Fibres and Textiles, 29*(3), 8–21. https://doi.org/10.15240/tul/008/2022-3-002

Duke University. (n.d.). *The Apparel Industry*. Available at: https://sites.duke.edu/sociol342d_01d_s2017_team-7/2-global-value-chain/. Accessed 2nd April 2024.

Dyck, A., Lins, K. V., Roth, L., & Wagner, H. F. (2019). Do institutional investors drive corporate social responsibility? International evidence. *Journal of Financial Economics, 131*(3), 693–714.

EPA. (2023). *Environmental Justice*. Online. Available at: https://www.epa.gov/environmentaljustice. Accessed 10th April 2024.

Eskeland, G. S., & Harrison, A. E. (2003). Moving to greener pastures? Multinationals and the pollution haven hypothesis. *Journal of Development Economics, 70*(1), 1–23.

Fei, W., Opoku, A., Agyekum, K., Oppon, J. A., Ahmed, V., Chen, C., & Lok, K. L. (2021). The critical role of the construction industry in achieving the sustainable development goals (SDGs): Delivering projects for the common good. *Sustainability*. https://doi.org/10.3390/su13169112

Felgueiras, C., Azoia, N. G., Gonçalves, C., Gama, M., & Dourado, F. (2021). Trends on the cellulose-based textiles: Raw materials and technologies. *Frontiers in Bioengineering and Biotechnology, 9*, 608826.

Fernandez-Stark, K., & Gereffi, G. (2019). Global value chain analysis: A primer. In S. Ponte, G. Gereffi, & G. Raj-Reichert (Eds.), *Handbook on global value chains* (pp. 54–76). Edward Elgar Publishing.

Fernandez-Stark, K., Frederick, S., & Gereffi, G. (2011). The apparel global value chain: Economic upgrading and workforce development. In G. Gereffi, K. Fernandez-Stark, & P. Psilos (Eds.), *Skills for upgrading: Workforce development and global value chains in developing countries*. Center on Globalization Governance & Competitiveness and RTI International.

Fernandez-Stark, K., Bamber, P. and Gereffi, G. (2013). Costa Rica in the offshore services global value chain.

Filho, W. L. (2012). World trends in education for sustainable development. https://doi.org/10.3726/978-3-653-04538-3

Francoeur, C., Lakhal, F., Gaaya, S., & Saad, I. B. (2021). How do powerful CEOs influence corporate environmental performance? *Economic Modelling, 94*, 121–129.

Gereffi, G. (2005). The global economy: Organization, governance, and development. *The Handbook of Economic Sociology, 2*, 160–182.

Gereffi, G. (2019). Global value chains and international development policy: Bringing firms, networks and policy-engaged scholarship back in. *Journal of International Business Policy, 2*(3), 195–210.

Gereffi, G., & Frederick, S. (2010). The global apparel value chain, trade and the crisis: Challenges and opportunities for developing countries.

Gereffi, G., & Lee, J. (2012). Why the world suddenly cares about global supply chains. *Journal of Supply Chain Management, 48*(3), 24–32.

Gereffi, G., Humphrey, J., & Sturgeon, T. (2005). The governance of global value chains. *Review of International Political Economy, 12*(1), 78–104.

Ghosh, B., & Mohan, M. A. P. (2021). Recent trends in sustainable textiles and apparel production. *International Journal of Current Science Research and Review, 4*(02), 78–81.

Goger, A. (2013). The making of a 'business case'for environmental upgrading: Sri Lanka's eco-factories. *Geoforum, 47*, 73–83.

Gölgeci, I., Gligor, D. M., Tatoglu, E., & Arda, O. A. (2019). A relational view of environmental performance: What role do environmental collaboration and cross-functional alignment play? *Journal of Business Research, 96*, 35–46. https://doi.org/10.1016/j.jbusres.2018.10.058

Golgeci, I., Makhmadshoev, D., & Demirbag, M. (2021). Global value chains and the environmental sustainability of emerging market firms: A systematic review of literature and research agenda. *International Business Review, 30*(5). https://doi.org/10.1016/j.ibusrev.2021.101857

Golini, R., De Marchi, V., Boffelli, A., & Kalchschmidt, M. (2018). Which governance structures drive economic, environmental, and social upgrading? A quantitative analysis in the assembly industries. *International Journal of Production Economics, 203*, 13–23.

Govindarajulu, N., & Daily, B. F. (2004). Motivating employees for environmental improvement. *Industrial Management & Data Systems, 104*(4), 364–372.

Goworek, H., Fisher, T., Cooper, T., Woodward, S., & Hiller, A. (2012). The sustainable clothing market: An evaluation of potential strategies for UK retailers. *International Journal of Retail & Distribution Management, 40*(12), 935–955.

Guan, C., & Qamruzzaman, M. (2022). A symmetric and asymmetric nexus between environmental sustainability and tourism development in BRIC nations: What is the role of good governance and globalization? *Frontiers in Environmental Science*. https://doi.org/10.3389/fenvs.2022.973420

Habib, A., & Bao, Y. (2019). Impact of knowledge management capability and green supply chain management practices on firm performance. *International Journal of Research in Business and Social Science (2147-4478), 8*(6), 240–255.

Habib, M. A., Balasubramanian, S., Shukla, V., Chitakunye, D., & Chanchaichujit, J. (2022). Practices and performance outcomes of green supply chain management initiatives in the garment industry. *Management of Environmental Quality an International Journal*. https://doi.org/10.1108/meq-08-2021-0189

Harari, Y. N. (2014). *Sapiens: A brief history of humankind*. Random House.

He, B., Chen, Y., Ismail, N., Chen, G., & Ni, L. (2024). The impact of digital empowerment on the economic openness of the textile industry in Keqiao based on a sustainable environment. *Heliyon, 10*(1).

Henninger, C. E., Alevizou, P. J., & Oates, C. J. (2016). What is sustainable fashion? *Journal of Fashion Marketing and Management: An International Journal, 20*(4), 400–416.

Herrigel, G., Wittke, V., & Voskamp, U. (2013). The process of C hinese manufacturing upgrading: Transitioning from unilateral to recursive mutual learning relations. *Global Strategy Journal, 3*(1), 109–125.

Hoang, D. P., Doan, N. T., & Nguyen, T. C. T. (2022). What motivates Vietnamese enterprises to upgrade in global value chains? Closing the intention-behavior gap. *Asia Pacific Journal of Marketing and Logistics, 34*(7), 1325–1347. https://doi.org/10.1108/apjml-06-2021-0392

Hollweg, C. H. (2019). *Global value chains and employment in developing economies*, Global Value Chain Development Report, 2019, p. 63.

Hoque, M. A., Rasiah, R., Furuoka, F., & Kumar, S. (2022). Critical determinants and firm performance of sustainable technology adoption in the apparel industry: The stakeholder approach. *Journal of Fashion Marketing and Management*. https://doi.org/10.1108/jfmm-06-2021-0147

Hou, F., Su, W., Cheng, S., Wu, C., & Lin, Y. (2023). The impact of environmental regulation on Hebei's manufacturing industry in the global value chain. *International Journal of Environmental Research and Public Health, 20*(4), 2933.

Humphrey, J., & Schmitz, H. (2002). How does insertion in global value chains affect upgrading in industrial clusters? *Regional Studies, 36*(9), 1017–1027. https://doi.org/10.1080/0034340022000022198

Huo, B. (2012). The impact of supply chain integration on company performance: An organizational capability perspective. *Supply Chain Management: An International Journal, 17*(6), 596–610.

Hussain, H. I., Haseeb, M., Kot, S., & Jermsittiparsert, K. (2020). Non-linear impact of textile and clothing manufacturing on economic growth: The case of top-Asian economies. *Fibres & Textiles in Eastern Europe, 5*(143), 27–36.

Iqbal, T., Shahzad, M. A., Alonso-Nuez, M. J., & Rosell-Martínez, J. (2022). Importance of environmental policy on firm performance for the textile industry: A contextual study of Pakistan. *Frontiers in Psychology, 13*, 1008890.

Islam, M. T., Jahan, R., Jahan, M., Howlader, M. S., Islam, R., Islam, M. M., Hossen, M. S., Kumar, A., & Robin, A. H. (2022). Sustainable textile industry: An overview. *Non-Metallic Material Science*. https://doi.org/10.30564/nmms.v4i2.4707

Jegethesan, K., Sneddon, J., & Soutar, G. N. (2012). Young Australian consumers' preferences for fashion apparel attributes. *Journal of Fashion Marketing and Management*. https://doi.org/10.1108/13612021211246044

Jensen, F., & Whitfield, L. (2022). Leveraging participation in apparel global supply chains through green industrialization strategies: Implications for low-income countries. *Ecological Economics, 194*. https://doi.org/10.1016/j.ecolecon.2021.107331

Jeppesen, S., & Hansen, M. W. (2004). Environmental upgrading of third world enterprises through linkages to transnational corporations. Theoretical perspectives and preliminary evidence. *Business Strategy and the Environment, 13*(4), 261–274.

Jia, F., Yin, S., Chen, L., & Chen, X. (2020). The circular economy in the textile and apparel industry: A systematic literature review. *Journal of Cleaner Production, 259*, 120728.

Jingbo, C., Shaozhou, Q., Jia, X., & Shu-Hua, Z. (2020). Institutional investors and corporate environmental innovation. *Sustainability at Work eJournal*. https://doi.org/10.2139/ssrn.3706637

Joy, A., Sherry, J. F., Jr., Venkatesh, A., Wang, J., & Chan, R. (2012). Fast fashion, sustainability, and the ethical appeal of luxury brands. *Fashion Theory, 16*(3), 273–295.

Juanga-Labayen, J. P., Labayen, I. V., & Yuan, Q. (2022). A review on textile recycling practices and challenges. *Text, 2*(1), 174–188. Available at: https://www.mdpi.com/2673-7248/2/1/10

Kadarusman, Y., & Nadvi, K. (2013). Competitiveness and technological upgrading in global value chains: Evidence from the Indonesian electronics and garment sectors. *European Planning Studies, 21*(7), 1007–1028. https://doi.org/10.1080/09654313.2013.733850

Kang, J., Liu, C., & Kim, S.-H. (2013). Environmentally sustainable textile and apparel consumption: The role of consumer knowledge, perceived consumer effectiveness and perceived personal relevance. *International Journal of Consumer Studies*. https://doi.org/10.1111/ijcs.12013

Kano, L. (2018). Global value chain governance: A relational perspective. *Journal of International Business Studies, 49*(6), 684–705. https://doi.org/10.1057/s41267-017-0086-8

Kar, S., & Kar, M. (2015). Multi-market firms and export quota: Effects of withdrawal of the multi-fiber arrangement. In *Understanding development: An Indian perspective on legal and economic policy* (pp. 99–113). Springer.

Karthik, T., & Gopalakrishnan, D. (2014). Environmental analysis of textile value chain: An overview. In S. S. Muthu (Ed.), *Roadmap to sustainable textiles and clothing: Environmental and social aspects of textiles and clothing supply chain* (pp. 153–188). Springer Singapore.

Khan, S. R., Dhar, D., Navaid, M., Pradhananga, M., Siddique, F., Singh, A., Yanthrawaduge, S., Khan, S. R., Dhar, D., & Navaid, M. (2009). 'The textile and readymade garment industry in south asia: A brief history and reemergence. In *Export success and industrial linkages: The case of readymade garments in South Asia* (pp. 25–40).

Khan, M. J., Ponte, S., & Lund-Thomsen, P. (2019). The 'factory manager dilemma': Purchasing practices and environmental upgrading in apparel global value chains. *Environment and Planning A: Economy and Space, 52*(4), 766–789. https://doi.org/10.1177/0308518x19876945

Khan, H., Weili, L., & Khan, I. (2021). Environmental innovation, trade openness and quality institutions: An integrated investigation about environmental sustainability. *Environment, Development and Sustainability, 24*(3), 3832–3862. https://doi.org/10.1007/s10668-021-01590-y

Khattak, A. (2013). *A comparative analysis of South Asian apparel firms in global value chains: Governance, institutions and upgrading*. ResearchSpace@ Auckland.

Khattak, A., & Park, Y.-E. (2018). Environmental upgrading of an apparel firm in Bangladesh: A case study of VIYELLATEX. *Emerald Emerging Markets Case Studies, 8*(3), 1–16. https://doi.org/10.1108/eemcs-06-2017-0129

Khattak, A., & Stringer, C. (2016). The role of suppliers in the greening of GVCs: Evidence from the sri lankan apparel industry. In M. Erdogdu, T. Arun, & I. Ahmad (Eds.), *Handbook of research on green economic development initiatives and strategies* (pp. 539–559). IGI Global.

Khattak, A., & Stringer, C. (2017). Environmental upgrading in Pakistan's sporting goods industry in global value chains: A question of Progress? *Business & Economic Review, 9*(1), 43–64.

Khattak, A., Stringer, C., Benson-Rea, M., & Haworth, N. (2015). Environmental upgrading of apparel firms in global value chains: Evidence from Sri Lanka. *Competition & Change, 19*(4), 317–335. https://doi.org/10.1177/1024529415581972

Ki, C. W., Park, S., & Ha-Brookshire, J. E. (2021). Toward a circular economy: Understanding consumers' moral stance on corporations' and individuals' responsibilities in creating a circular fashion economy. *Business Strategy and the Environment, 30*(2), 1121–1135.

Kim, J. O., Traore, M. K., & Warfield, C. (2006). The textile and apparel industry in developing countries. *Textile Progress, 38*(3), 1–64. https://doi.org/10.1533/tepr.2006.0003

Köksal, D., Strähle, J., Müller, M., & Freise, M. (2017). Social sustainable supply chain management in the textile and apparel industry—A literature review. *Sustainability*. https://doi.org/10.3390/su9010100

Koltonuk, D. J. (2020). The impact of environmental sustainability on venture capital, private equity, and investor relations: Exploring Lapland, Finland as a space for sustainable investment.

Krauss, J. E., & Krishnan, A. (2021). Global decisions versus local realities: Sustainability standards, priorities and upgrading dynamics in agricultural global production networks. *Global Networks, 22*(1), 65–88. https://doi.org/10.1111/glob.12325

Krishnan, A., De Marchi, V., & Ponte, S. (2023). Environmental upgrading and downgrading in global value chains: A framework for analysis. *Economic Geography, 99*(1), 25–50. https://doi.org/10.1080/00130095.2022.2100340

Kummritz, V., Taglioni, D., & Winkler, D. E. (2017) *Economic upgrading through global value chain participation: Which policies increase the value added gains?*, World Bank Policy Research Working Paper, (8007).

La Rosa, A. D., & Grammatikos, S. A. (2019). Comparative life cycle assessment of cotton and other natural fibers for textile applications. *Fibers, 7*(12), 101. Available at: https://www.mdpi.com/2079-6439/7/12/101

Laitala, K., Austgulen, M. H., & Klepp, I. G. (2014). Responsibility without means. In S. S. Muthu (Ed.), *Roadmap to sustainable textiles and clothing: Environmental and social aspects of textiles and clothing supply chain* (pp. 125–151). Springer Singapore.

Langford, N. J., Nadvi, K., & Braun-Munzinger, C. (2023). The shaping of 'Southern' sustainability standards in a value chain world: Comparative evidence from China and India. *Review of International Political Economy, 30*(3), 1151–1176. https://doi.org/10.1080/09692290.2022.2089713

Laosirihongthong, T., Samaranayake, P., & Nagalingam, S. V. (2019). A holistic approach to supplier evaluation and order allocation towards sustainable procurement. *Benchmarking an International Journal*. https://doi.org/10.1108/bij-11-2018-0360

Lechner, C., Lorenzoni, G., Guercini, S., & Gueguen, G. (2020). Supplier evolution in global value chains and the new brand game from an attention-based view. *Global Strategy Journal, 10*(3), 520–555.

Lee, S.-Y. (2016). Responsible supply chain management in the Asian context: The effects on relationship commitment and supplier performance. *Asia Pacific Business Review, 22*(2), 325–342.

Lewis, T., Zeng, X., Sanchez, V., & Fan, J. (2017). Environmental evaluation of fabric dyeing and water use for a global apparel manufacturer. *Aatcc Journal of Research*. https://doi.org/10.14504/ajr.4.1.1

Li, J., & Leonas, K. K. (2021). The impact of communication on consumer knowledge of environmentally sustainable apparel. *Journal of Fashion Marketing and Management*. https://doi.org/10.1108/jfmm-02-2021-0034

Li, Y., Wang, Y., Zhang, X., & Huang, Q. (2022). Impact of global value chain embedding on industrial environmental performance: An empirical study based on the countries along the "Belt and Road". *Frontiers in Environmental Science, 10*. https://doi.org/10.3389/fenvs.2022.1026068

Li, L., Liu, X., & Hu, M. (2024). Textile and apparel supply chain coordination under ESG related cost-sharing contract based on stochastic demand. *Journal of Cleaner Production, 437*, 140491.

Lister, J., Poulsen, R. T., & Ponte, S. (2015). Orchestrating transnational environmental governance in maritime shipping. *Global Environmental Change, 34*, 185–195.

Lopez-Acevedo, G., & Robertson, R. (2016). Stitches to riches?: Apparel employment, trade, and economic development in South Asia. https://doi.org/10.1596/978-1-4648-0813-5

Luken, R., & Stares, R. (2005). Small business responsibility in developing countries: A threat or an opportunity? *Business Strategy and the Environment, 14*(1), 38–53.

Lund-Thomsen, P. (2004). Towards a critical framework on corporate social and environmental responsibility in the South: The case of Pakistan. *Development, 47*(3), 106–113.

Mahmood, H., Tanveer, M., & Furqan, M. (2021). 'Rule of law, corruption control, governance, and economic growth in managing renewable and nonrenewable energy consumption in South Asia. *International Journal of Environmental Research and Public Health*. https://doi.org/10.3390/ijerph182010637

Mahmoudi, R., & Rasti-Barzoki, M. (2018). Sustainable supply chains under government intervention with a real-world case study: An evolutionary game theoretic approach. *Computers & Industrial Engineering, 116*, 130–143.

Marques, A. D., Marques, A., & Ferreira, F. (2020). Homo Sustentabilis: Circular economy and new business models in fashion industry. *SN Applied Sciences*. https://doi.org/10.1007/s42452-020-2094-8

Mastamet-Mason, A., & More, C. J. (2023). Perceived environmental implications of clothing maintenance among consumers in Gauteng Province, South Africa. *Academic Journal of Interdisciplinary Studies*. https://doi.org/10.36941/ajis-2023-0035

Meng, S., Yan, H., & Yu, J. (2022). Global value chain participation and green innovation: Evidence from Chinese listed firms. *International Journal of Environmental Research and Public Health, 19*(14), 8403.

Milberg, W., & Winkler, D. (2011). Economic and social upgrading in global production networks: Problems of theory and measurement. *International Labour Review, 150*(3–4), 341–365. https://doi.org/10.1111/j.1564-913X.2011.00120.x

Milberg, W., & Winkler, D. (2013). *Outsourcing economics: Global value chains in capitalist development*. Cambridge University Press.

Mohai, P., Pellow, D., & Roberts, J. T. (2009). Environmental justice. *Annual Review of Environment and Resources, 34*, 405–430. https://doi.org/10.1146/annurev-environ-082508-094348

Moreau, V., Sahakian, M., Van Griethuysen, P., & Vuille, F. (2017). Coming full circle: Why social and institutional dimensions matter for the circular economy. *Journal of Industrial Ecology, 21*(3), 497–506.

Nand, A. A., Menon, R., Bhattacharya, A., & Bhamra, R. (2022). A review of sustainability trade-offs affecting suppliers in developed and less developed countries. *Journal of Business & Industrial Marketing, 38*(3), 463–483. https://doi.org/10.1108/jbim-04-2021-0213

Navarrete, S. D. S., Borini, F. M., & Avrichir, I. (2020). Environmental upgrading and the United Nations Sustainable Development Goals. *Journal of Cleaner Production, 264*. https://doi.org/10.1016/j.jclepro.2020.121563

Nguyen, N. H., Beeton, R. J. S., & Halog, A. (2014). Who influence the environmental adaptation process of small and medium sized textile and garment companies in Vietnam? In S. S. Muthu (Ed.), *Roadmap to sustainable textiles and clothing: Environmental and social aspects of textiles and clothing supply chain* (pp. 189–207). Springer Singapore.

Niinimäki, K., Peters, G., Dahlbo, H., Perry, P., Rissanen, T., & Gwilt, A. (2020). The environmental price of fast fashion. *Nature Reviews Earth & Environment, 1*(4), 189–200.

Nikmatin, S., Hermawan, B., Kardiansyah, T., Seta, F. T., Afiah, I. N., & Misbah, M. (2022). Oil palm empty fruit bunches as raw material of dissolving pulp for viscose rayon fiber in making textile products. *Polymers*. https://doi.org/10.3390/polym14153208

Orsato, R. J. (2006). Competitive environmental strategies: When does it pay to be green? *California Management Review, 48*(2), 127–143.

Pahl, S., & Timmer, M. P. (2020). Do global value chains enhance economic upgrading? A long view. *The Journal of Development Studies, 56*(9), 1683–1705.

Papú Carrone, N. (2020). Traceability and transparency: A way forward for SDG 12 in the textile and clothing industry. In *The UN Sustainable Development Goals for the textile and fashion industry*, pp. 1–19.

Peng, Y., Zhang, X., van Donk, D. P., & Wang, C. (2022). How can suppliers increase their buyers' CSR engagement: The role of internal and relational factors. *International Journal of Operations & Production Management, 42*(2), 206–229. https://doi.org/10.1108/ijopm-06-2021-0387

Ponte, S. (2021). Sustainability and green capital accumulation. In *Structural Transformation in South Africa: The Challenges of Inclusive Industrial Development in a Middle-Income Country*, pp. 143.

Ponte, S. (2022). The hidden costs of environmental upgrading in global value chains. *Review of International Political Economy, 29*(3), 818–843. https://doi.org/10.1080/09692290.2020.1816199

Ponte, S., De Marchi, V., Bettiol, M., & Di Maria, E. (2023). The horizontal governance of environmental upgrading: Lessons from the Prosecco and Valpolicella wine value chains in Italy. *Environment and Planning A: Economy and Space*, 0308518X231176553.

Porter, M. E., & Kramer, M. R. (2006). The link between competitive advantage and corporate social responsibility. *Harvard Business Review, 84*(12), 78–92.

Poulsen, R. T., Ponte, S., & Lister, J. (2016). Buyer-driven greening? Cargo-owners and environmental upgrading in maritime shipping. *Geoforum, 68*, 57–68. https://doi.org/10.1016/j.geoforum.2015.11.018

Rapsikevičienė, J., Gurauskienė, I., & Jučienė, A. (2019). Model of industrial textile waste management. *Environmental Research Engineering and Management*. https://doi.org/10.5755/j01.erem.75.1.21703

Raut, R., Gardas, B. B., & Narkhede, B. (2019). Ranking the barriers of sustainable textile and apparel supply chains: An interpretive structural modelling methodology. *Benchmarking: An International Journal, 26*(2), 371–394.

Reed, D. (2002). Resource extraction industries in developing countries. *Journal of Business Ethics, 39*(3), 199–226. https://doi.org/10.1023/A:1016538006160

Reis, A., Coelho, L., María, R., Batista, P., de Carvalho, M. R., Moura, B., Fernandes, S., Malgueiro, R., Pintado, M., da Ana Sofia Beça Ribas, S., Silva, M. J., Marques, M. M., Costa, C., Sá, V. L., Merino, F. M., Abreu, J. S., Marinho, A., Magalhães, A. I., Pedro, F., Magalhães, P. J., Silva, R. J., Pereira, J. A., Gaião, J. M., Vilaça, H., Ventura, S., Teles, Â., Mourão, R. R., Morgado, J. C., & Gonçalves, M. S. (2020). Innovation of textiles through natural by-products and wastes. *DOI*. https://doi.org/10.5772/intechopen.93011

Reps, N., & Braun, B. (2012). Going green - environmental upgrading and value chain coordination in the Indian automotive industry. *Zeitschrift Fur Wirtschaftsgeographie, 56*(4), 226–243. Available at: <Go to ISI>://WOS:000314200400002.

Rossi, A. (2019). *Social upgrading* (pp. 272–284). Edward Elgar Publishing.

Schmitz, H. (2006). Learning and earning in global garment and footwear chains. *The European Journal of Development Research, 18*, 546–571.

Shen, B., Li, Q., Dong, C., & Perry, P. (2017). Sustainability issues in textile and apparel supply chains. *Sustainability*. https://doi.org/10.3390/su9091592

Shirvanimoghaddam, K., Motamed, B., Ramakrishna, S., & Naebe, M. (2020). Death by waste: Fashion and textile circular economy case. *Science of the Total Environment, 718*, 137317.

Sidian, Y., Ghazali, R., & Tajuddin, R. M. (2023). Integrating sustainable concepts into textile design courses: An effective teaching practices. *International Journal of Global Optimization and Its Application*. https://doi.org/10.56225/ijgoia.v2i1.159

Singh, S., & Jana, P. (2023). Decoding the science behind the chemical recycling of textiles'. In *International Conference on Functional Textiles & Clothing* (pp. 295–320). Springer.

Sinha, A., Mishra, S., Sharif, A., & Yarovaya, L. (2021). Does green financing help to improve environmental & social responsibility? Designing SDG framework through advanced quantile modelling. *Journal of Environmental Management, 292*, 112751.

Sinkovics, N., & Sinkovics, R. R. (2019). International business and global value chains. In S. Ponte, G. Gereffi, & G. Raj-Reichert (Eds.), *Handbook on global value chains* (pp. 417–431). Edward Elgar Publishing Limited.

Sinkovics, N., Sinkovics, R. R., Hoque, S. F., & Alford, M. (2017). The use of global value chain/global production network related literature in international business research: Investigating the nature and degree of integration. In G. Cook, J. Johns, F. McDonald, J. Beaverstock, & N. Pandit (Eds.), *The Routledge companion to the geography of international business* (pp. 258–296). Routledge.

Smith, D. A. (1996). Going south: Global restructuring and garment production in three East Asian cases. *Asian Perspective*, 211–241.

Su, J. (2013). Strategic sourcing in the textile and apparel industry. *Industrial Management & Data Systems*. https://doi.org/10.1108/02635571311289647

Suarez-Visbal, L. J., Rosales-Carreón, J., Corona, B., Hoffman, J., & Worrell, E. (2024). Transformative circular futures in the textile and apparel value chain: Guiding policy and business recommendations in The Netherlands, Spain, and India. *Journal of Cleaner Production*, 141512.

Sustainable Apparel Forum. (2023). Available at: https://sustainableapparelforum.com/. Accessed 10th Apr 2024.

Temminck, E., Mearns, K., & Fruhen, L. (2015). Motivating employees towards sustainable behaviour. *Business Strategy and the Environment, 24*(6), 402–412.

Tewari, M. (2006). Adjustment in India's textile and apparel industry: Reworking historical legacies in a post-Mfa world. *Environment and Planning a Economy and Space*. https://doi.org/10.1068/a38279

Thinakaran, S., Chandravelu, P., Ponnambalam, S. G., Sankaranarayanan, B., & Karuppiah, K. (2023). Analyzing the challenges to circular economy in Indian fashion industry. *IEEE Access*. https://doi.org/10.1109/access.2022.3233197

Tokatli, N. (2013). Toward a better understanding of the apparel industry: A critique of the upgrading literature. *Journal of Economic Geography, 13*(6), 993–1011.

Tokatli, N., Wrigley, N., & Kizilgün, Ö. (2008). Shifting global supply networks and fast fashion: Made in Turkey for Marks & Spencer. *Global Networks, 8*(3), 261–280.

Uddin, M. H., Razzak, M. R., & Rahman, A. A. (2023). Sustainable supply chain management practices, dynamic capabilities and competitive advantage: Evidence from Bangladesh <scp>ready-made</Scp> garments industry. *Business Strategy & Development*. https://doi.org/10.1002/bsd2.232

Vachon, S., & Klassen, R. D. (2008). Environmental management and manufacturing performance: The role of collaboration in the supply chain. *International Journal of Production Economics, 111*(2), 299–315.

Van Assche, A., & Narula, R. (2023). Internalization strikes back? Global value chains, and the rising costs of effective cascading compliance. *Journal of Industrial and Business Economics, 50*(1), 161–173. https://doi.org/10.1007/s40812-022-00237-x

Van Assche, A., Pasha, A. A., Cernat, L., & Voss, H. (2023). From the editor—Governments as buyers: The international business implications of public procurement. *Journal of International Business Policy*. https://doi.org/10.1057/s42214-023-00174-z

Vićentijević, K., Rakić, S., & Marković, V. (2022). Integration of CSR and SDG in creating added value of the textile industry. *Tekstilna industrija, 70*(3), 26–31.

Villanueva, I. D., & Lyu, S. (2020). Salvaging the value of textile waste through scratchboard etching. DOI: https://doi.org/10.31274/itaa.11969

Waite, M. (2014). The SURF framework applied to the textile industry. In S. S. Muthu (Ed.), *Roadmap to sustainable textiles and clothing: Environmental and social aspects of textiles and clothing supply chain* (pp. 209–226). Springer Singapore.

Watson, K. J., & Wiedemann, S. (2019). Review of methodological choices in LCA-based textile and apparel rating tools: Key issues and recommendations relating to assessment of fabrics made from natural fibre types. *Sustainability*. https://doi.org/10.3390/su11143846

Wilhelm, M., Blome, C., Wieck, E., & Xiao, C. (2016). Implementing sustainability in multi-tier supply chains: Strategies and contingencies in managing sub-suppliers. *International Journal of Production Economics*. https://doi.org/10.1016/j.ijpe.2016.08.006

Winakor, G. (1969). The process of clothing consumption. *Journal of Home Economics, 61*(8), 629–634.

Xiang-mei, Z., Zhang, B., & Yuan, H. (2022). Digital economy, industrial structure upgrading and green Total factor productivity—Evidence in textile and apparel industry from China. *PLoS One*. https://doi.org/10.1371/journal.pone.0277259

Xu, J., Liu, Y., & Yang, L. (2018). A comparative study of the role of China and India in sustainable textile competition in the U.S. market under green trade barriers. *Sustainability, 10*(5), 1348. Available at: https://www.mdpi.com/2071-1050/10/5/1348

Yao, L., & Geng, A.-Y. (2015). The environmental issues in textile and garment trade. In *Low-carbon City and New-type Urbanization: Proceedings of Chinese Low-carbon City Development International Conference* (pp. 121–128). Springer.

Yize, H., Jun, S., & Peixun, Z. (2020). Institutional investors' corporate site visits and firms' sustainable development. *Sustainability*. https://doi.org/10.3390/su12177036

Zhang, X., & Chen, Y. (2019). Carbon emission evaluation based on multi-objective balance of sewing assembly line in apparel industry. *Energies*. https://doi.org/10.3390/en12142783

Zhang, Q., & Hathcote, J. M. (2008). Factors influencing apparel imports from China. *Clothing and Textiles Research Journal, 26*(1), 23–40. https://doi.org/10.1177/0887302x07305922

Zhou, Y., Zhou, R., Chen, L., Zhao, Y., & Zhang, Q. (2022). Environmental policy mixes and green industrial development: An empirical study of the Chinese textile industry from 1998 to 2012. *Ieee Transactions on Engineering Management*. https://doi.org/10.1109/tem.2020.3009282

Analysis of Challenges to the Circular Economy Model of Textile Supply Chain Management Using TOE-TBL Approach

Emel Yontar

1 Introduction

It is an undeniable fact that the application of the circular economy model for the textile industry, which is known to be resource-intensive, will enable the industry to develop economically and have a positive environmental impact (John & Mishra, 2023). Because the textile industry creates significant environmental impacts due to the excessive use of natural resources, significant water use (Chen et al., 2017), and various chemicals (Al-Tohamy et al., 2022; Palacios-Mateo et al., 2021). Consumption of 98 million tons of non-renewable resources (oil, fertilizer, etc.) every year (Ellen MacArthur Foundation, 2017); since 2% greenhouse gas emission is generated worldwide (Sadowski et al., 2021) and only 20% of the produced textile product is recycled (Koszewska, 2018), the impact on climate change is 8% (Quintás et al., 2018). It also faces massive environmental and resource problems due to excessive waste from the fast fashion phenomenon (Koszewska, 2018). In addition to its environmental effects, it also has positive effects in economic and social terms. It provides more than 300 million jobs in low- and middle-income countries along the value chain (Ellen MacArthur Foundation, 2017) and plays an important role in the national economy by contributing to the national budget (Poursoltan et al., 2021). Higher living standards along with the increasing population bring the policy of being environmentalist in textile production to the agenda, and environmental transformations are taking place in the supply chains. The concept of the circular

E. Yontar (✉)
Center for Transportation & Logistics, Massachusetts Institute of Technology, Cambridge, MA, USA

Department of Industrial Engineering, Faculty of Engineering, Tarsus University, Mersin, Turkey
e-mail: eyontar@mit.edu; eyontar@tarsus.edu.tr

S. S. Muthu (ed.), *Sustainable Textile and Apparel Chain Management*, Sustainable Textiles: Production, Processing, Manufacturing & Chemistry,
https://doi.org/10.1007/978-3-031-80240-9_3

economy, which is necessary to transform this effect into a positive one, has been frequently discussed as a subject of study in recent years. Linking the textile industry with the circular economy can reduce the consumption of raw materials and prevent pollution during and after the production process (Filho Leal et al., 2019). Despite its increasing interest, the obstacles in front of it attract a lot of attention (Shirvanimoghaddam et al., 2020), and it is desired to be solved.

The circular economy is a model designed as opposed to the linear economy with its efficient use of natural resources, minimization of waste, and its compatibility with the sustainability system (Ranta et al., 2018). In the traditional linear economy model, resources are extracted and products are produced, used, and then disposed of. However, in the circular economy model, these resources are recycled or reused as much as possible (Ellen MacArthur Foundation, 2017). In other words, it adopts the 3R strategy (Reduce-Reuse-Recycle). The circular economy aims to use materials as efficiently as possible in the production, use, and recycling stages. In this way, it reduces the consumption of natural resources and minimizes the generation of waste and environmental impact. The circular economy contributes to sustainable production and consumption by reproducing and reusing discarded products (Ozkan-Ozen et al., 2020).

For the textile industry to have a positive environmental impact, there are some studies in the literature that adopt the circular economy model and identify it with the textile industry. In studies for textile sectors in different regions, Saccani et al. (2023), in their study, investigated the responses to the circular economy adoption barriers in supply chain management for the textile and fashion industry in the Italian wool region and shared information about the circular supply chain. Galatti and Baruque-Ramos (2022), addressing social innovation, 40 evaluation indicators of the Brazilian textile and fashion industry that are likely to promote social innovation in the circular economy and determined the leading criteria for employment, occupational health and safety, and labor relations. Fischer and Pascucci (2017) investigated the promotion of the necessary conditions for the transition to the circular economy in inter-firm cooperation in the textile industry in the Netherlands and proposed a conceptual framework for these studies. Balanay and Halog (2019) proposed a roadmap for the country to provide a circular model economy in textile industry activities in the Philippines. Huang et al. (2021) explored overcoming circular economy barriers to foster innovation in the textile industry in Taiwan. It was analyzed with Fuzzy Delphi Method, Total interpretative structural modeling, and MICMAC methods. As a result of the analysis, low customer demand for recycled textiles, difficulty in collaborative innovation among supply chain partners, lack of high-quality recycling materials, lack of successful circular business models, and high costs but low economic benefits in the short term emerged as the most important challenges. Akter et al. (2022) investigated the circular economy model to achieve sustainable development goals for textile material waste management in Bangladesh. In addition to the studies carried out in these different regions, Majumdar et al. (2022), again focusing on the obstacles, aimed to identify the obstacles in front of the circular model in the textile and clothing supply chain. Eighteen barriers were addressed, and using the Grey-DEMATEL analysis, lack of supply chain cooperation, lack of effective planning, lack of incentives, and lack of technical knowledge for recycling emerged as major barriers. Salmi and Kaipia (2022), in their studies, focused on the challenges and the transformation process in order to overcome the past studies in the textile sector for the transition to the circular economy model. Luoma et al. (2022), by addressing the concept of digitalization, explained that the circular economy is a driving force with an example in the textile industry and emphasized that the transition is difficult.

There are also studies focusing on products and production processes. Franco (2017) identified factors for a firm's desire to develop a circular product at all stages of product design along the textile value chain, drawing attention to the increase in circular economy studies in the textile sector, but its lack in terms of implementation. Likewise, Teixeira et al. (2023) aimed to determine green innovation and circular economy practices in the new product development process in the textile industry. In addition to product-oriented studies, Neto de Oliveira et al. (2022), who innovated in the production process, aimed to evaluate the degree of sustainability according to the size of the textile sector in circular economy studies by targeting clean production practices and concluded this with ANOVA analysis. Silva and Morais (2022) proposed a new multi-criteria decision model for the circular economy model to separate outsourcing for the textile industry determined by a decision support system. Jäämaa and Kaipia (2022) suggested in their study how the collection of end-of-life textile products from households with first-mile problem can be modeled in a situation where historical volume data is not available. John and Mishra (2023) developed a sustainable circular three-tier supply chain model for the textile and apparel industries, consisting of a single supplier and manufacturer and multiple retailers.

In addition to these studies, there are studies that carry out a literature review; Jia et al. (2020) conducted a literature review including practices, driving forces, obstacles, and indicators by considering sustainability performance in the textile industry from a circular economy perspective. In addition, the relationship within these four themes is detailed with the conceptual model. Abdelmeguid et al. (2022), who conducted a similar study using PRISMA analysis, conducted a systematic literature review to summarize the challenges of implementing the circular economy in the fashion industry.

Despite the well-known net benefits, the transition from linear economy to circular economy brings many difficulties and obstacles (Kayikci et al., 2022). Researchers have generally discussed the circular economy with different types, such as financial, social, and infrastructure (Kumar et al., 2019; Vermunt et al., 2019), and the barriers to the circular economy model in relation to the textile sector have been investigated (Jia et al., 2020; Koszewska, 2018; Majumdar et al., 2022).

As a result of a detailed review of the literature, it has been determined that the sustainability-oriented TOE-TBL approach, which will contribute to the textile sector as well as other sectors, is not applied to circular economy obstacles when every stage of supply chain management is taken into consideration, and the primary contribution of the study to the literature is the evaluation of obstacles with the TOE-TBL approach within the framework of circular economy.

In the current study for the studies in the literature, it is aimed to investigate the challenges encountered in the supply chain management of the circular economy model in the textile sector from raw material purchase to the production stage, from distribution to after-sales textile products recycling stage. For this purpose, it is aimed to seek answers to the following questions:

RQ1: What are the criteria that prevent the application of the circular economy model in the supply chain management stages in the textile industry?
RQ2: Is it possible to classify these identified challenges among the three dimensions of sustainability from the perspective of technology, organization, and environment?
RQ3: Does prioritizing circular economy challenges with different multi-criteria decision making (MCDM) methods identify possible dominant challenges?

Based on a comprehensive literature review and input from experts working in the textile industry, this research answers the first question by listing the barriers to the transition to a circular economy. The three dimensions of sustainability, economic, social and environmental main factors, and the classification of the obstacles listed with the technology-organization-environment approach answers the second question. Finally, the evaluation of the obstacles obtained by the decision makers with SWARA (Step-Wise Weight Assessment Ratio Analysis)-AHP (Analytic Hierarchy Process)-DEMATEL (The Decision Making Trial and Evaluation Laboratory)-COPRAS (Complex Proportional Assessment)-TOPSIS (Technique for Order Preference by Similarity to Ideal Solutions) methods, which is one of the different MCDM techniques, explaining the relationship between them and presenting the comparison answers the last question.

Other parts of the study continue as follows: In Sect. 2, the conceptual model of the study is given, and the flow of the study is explained. MCDM techniques are theoretically described in this section. Section 3 includes the implementation phase of the study and deals with the determination of the criteria and then their prioritization and the impact assessment of the criteria themselves. In Sect. 4, the place of barriers in the TOE-TBL approach is explained, and in Sect. 5, theoretical and practical implication is mentioned. Also, the study is concluded by giving place to the discussion in Sect. 6.

2 Methodology

In this section of the study, the conceptual model of the chapter (Fig. 1) is given, and explanations are made about MCDM methods. With MCDM methods, tools that give possible robust and powerful results can be used in critical decision-making processes. In this study, considering every stage of textile supply chain management, the obstacles to the transition to the circular economy in the literature, and as a result of group discussions with experts, circular economy obstacles were listed in the focus of TOE-TBL and evaluated with MCDM techniques.

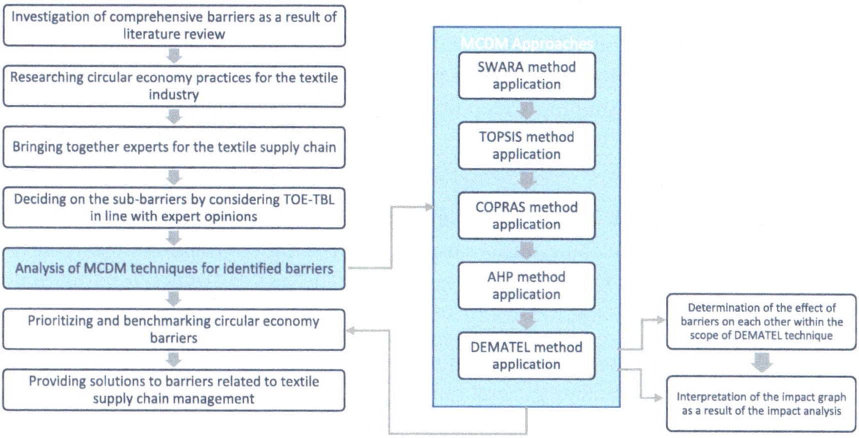

Fig. 1 The conceptual model

2.1 *MCDM Approaches*

In the decision-making process, there are many factors that need to be considered for each problem. While the number of factors varies according to the nature of the problem, the decision maker considers the most important of these factors as criteria. Because the number of criteria to include in MCDM methods is a complex issue (Villacreses et al., 2022). On the other hand, there is more than one purpose for making a choice in decision-making problems. Various stakeholders, experts, and researchers use MCDM techniques to decide and evaluate numerous criteria for these purposes (Abishini & Karthikeyan, 2023). It brings with it many difficulties, and it is requested by researchers and experts to analyze it in limited information under uncertainty. Over the course of time, various methods have been developed to overcome these difficulties, and the focus has always been on achieving the goal. This study aims to explore the circular economy barriers for the textile industry and to rank their importance with SWARA, AHP, TOPSIS, COPRAS, and DEMATEL techniques, considering the different aspects of MCDM techniques. In most MCDM techniques in the literature, a scoring between alternatives and criteria is requested by the decision maker and converted into a matrix. The following two steps constitute the first step of the techniques, except SWARA. Customized steps are included below each technique.

Step 1: Creating the decision matrix. The decision matrix consists of an initial decision matrix with m alternatives and n criteria created by the decision maker.

Step 2: Creation of the standard decision matrix. Normalization or standardization process is performed so that the alternatives evaluated in different scales can be brought to the same scale and compared.

Although SWARA, AHP, TOPSIS, COPRAS, and DEMATEL techniques are considered to be powerful and valid methods with their effective algorithms, they have the possibility to produce different rankings for the criteria. In this regard, the results were compared using these techniques to further increase the reliability of the findings.

SWARA

The SWARA (Step-Wise Weight Assessment Ratio Analysis) method is a method in which expert opinions can be easily included in the process and simple relative comparison can be used (Alimardani et al., 2013). The process of determining their relative weights includes the following steps (Stanujkic et al., 2015):

Step 1: The criteria are sorted in descending order of importance according to expert opinion. In order to obtain a general ranking when there is more than one decision maker, the geometric average of these operations is taken.

Step 2: The relative importance of each criterion is determined. For this *J*. criterion, ($j + 1$). by comparison with the criteria *j*. your criterion ($j + 1$). how important it is is determined from the criterion (Ruzgys et al., 2014). This value is shown as s_j (comparative significance of the mean value) (Keršuliene et al., 2010).

Step 3: The coefficient k_j is determined as in (1).

$$kj = \left\{ \begin{matrix} 1 \\ sj + 1 \end{matrix} \Big| \begin{matrix} j=1 \\ j>1 \end{matrix} \right\} \tag{1}$$

Step 4: q_j variable is calculated as in (2).

$$qj = \left\{ \begin{matrix} 1 \\ \dfrac{qj-1}{kj} \end{matrix} \Big| \begin{matrix} j=1 \\ j>1 \end{matrix} \right\} \tag{2}$$

Step 5: The relative weights of the evaluation criteria are determined by the formula in Eq. (3). Here, w_j, *j*. specifies the relative weight of the criterion.

$$wj = \frac{qj}{\sum_{k=1}^{n} qk} \tag{3}$$

AHP

AHP (Analytic Hierarchy Process) determines the set of criteria and the relative importance of these criteria based on expert evaluations. Thus, quantitative performance measurements are combined with subjective evaluations with a systematic approach to obtain healthy results (Tektas & Alakavuk, 2003). Decision steps are listed as follows (Saaty, 1988):

Step 1: Defining the decision problem.

Step 2: Establishing the hierarchical structure.

Step 3: Creating pairwise comparison matrices. In AHP, the relative importance scale developed by Saaty is used to measure and create pairwise comparison matrices.

Step 4: Conversion to vector priorities. The comparison matrix shows the importance levels of the factors relative to each other in a certain logic. However, in order to determine the weights of these factors in the whole, in other words, the percentage importance distributions, the column vectors that make up the comparison matrix are used and a B column vector with n and n components is created (4). When n B column vectors are combined in a matrix format, the C matrix shown in (6) will be formed (Yaralıoğlu, 2001).

$$B_i = \begin{bmatrix} b_{1i} \\ b_{2i} \\ \vdots \\ b_{ni} \end{bmatrix} \tag{4}$$

$$b_{ij} = \frac{a_{ij}}{\sum\limits_{i=1}^{n} a_{ij}} \tag{5}$$

$$C = \begin{bmatrix} b_{11} & b_{12} & \cdots & b_{1n} \\ b_{21} & b_{22} & \cdots & b_{2n} \\ \cdots & \cdots & \ddots & \cdots \\ b_{n1} & b_{n2} & \cdots & b_{nn} \end{bmatrix} \tag{6}$$

By using the *C* matrix, the percent importance distributions showing the relative importance values of the factors can be obtained. For this, the arithmetic average of the row components forming the *C* matrix is taken, and the *W* column vector called the Priority Vector is obtained (7) and (8).

$$W_i = \frac{\sum\limits_{j=1}^{n} b_{ij}}{n} \tag{7}$$

$$W = \begin{bmatrix} w_1 \\ w_2 \\ \vdots \\ w_n \end{bmatrix} \quad (8)$$

Thus, in order to achieve the goal, the priorities of the elements, in other words, the relative importance of each criterion and the relative importance of each decision alternative with respect to the relevant criterion, are determined.

Step 5: Calculating concordance rate. Consistency ratio, abbreviated as CR, is used to measure whether the decision maker behaves consistently when comparing criteria.

Step 6: Sorting the options.

TOPSIS

The TOPSIS (Technique for Order Preference by Similarity to Ideal Solutions) method is based on the idea that the best alternative chosen should have the alternative closest to the positive ideal solution and the farthest from the negative ideal solution (Gong & Shi, 2008).

Step 1: Creating the weighted decision matrix. In this step, the weights of the evaluation factors are determined so that the sum of the weights is 1. These determined weights are expressed as w_j. After the weights are determined, each element of the normalized decision matrix (r_{ij}) is multiplied by the weight of the relevant criterion (w_j).

Step 2: Generating positive and negative ideal solutions. In order to reach the ideal solution (A^*), the largest and smallest values in each column of the weighted decision matrix created in Step 1 are selected. The smallest values are clustered as A^- as seen in (9), and the largest values as A^+ as seen in (10). The points A^- indicate the negative ideal point, and the points A^+ indicate the positive ideal point.

$$A^- = \left\{ \left(\min{}_i v_{ij} | j \in J \right), \left(\max{}_i v_{ij} | j \in J' \right) \right\} \quad (9)$$

$$A^+ = \left\{ \left(\max{}_i v_{ij} | j \in J \right), \ \left(\min{}_i v_{ij} | j \in J' \right) \right\} \quad (10)$$

Step 3: Calculation of the separation measures. After the ideal points are defined, the distance values to the maximum and minimum ideal points are calculated with the help of the following formulas:

$$S_i^+ = \sqrt{\sum_{j=1}^{m} \left(v_{ij} - v_j^+\right)^2} \tag{11}$$

$$S_i^- = \sqrt{\sum_{j=1}^{m} \left(v_{ij} - v_j^-\right)^2} \tag{12}$$

Step 4: Calculating the relative closeness to the ideal solution. While calculating the closeness values with respect to the ideal point for each alternative, the relative closeness values to the ideal and negative ideal points are used.

$$C_i^* = \frac{S_i^-}{S_i^- + S_i^*} \tag{13}$$

$$0 \leq C_i^* \leq 1 \tag{14}$$

C_i^*=1 shows the absolute closeness of the relevant decision point to the ideal solution, and C_i^*=0 shows the absolute closeness of the relevant decision point to the negative ideal solution. Finally, the values obtained are arranged in order of magnitude and the order of importance of the decision points (alternatives) is determined.

COPRAS

The COPRAS (Complex Proportional Assessment) method, which can handle both quantitative and qualitative criteria, is a method that allows the full ranking of decision alternatives to be obtained (Ayçin, 2020).

Step 4: Collecting the weighted normalized indexes. The smaller the $S\!-\!j$ value, the higher the goal is to reach. The greater the value of $S\!+\!j$, the higher the achievement of the goal (15).

$$S_{+j} = \sum_{i=1}^{m} d + ij; \qquad S_{-j} = \sum_{i=1}^{m} d - ij \tag{15}$$

Step 5: Calculating the relative importance of decision alternatives. *Qj*, which shows the relative importance of the options being compared, is ordered from largest to smallest.

$$Q_j = S_{+j} + \frac{S_{-\min} \sum_{j=1}^{n} S_{-j}}{S_{-\min} \sum_{j=1}^{n} \frac{S_{-\min}}{S_{-j}}} \tag{16}$$

Step 6: Calculation of the performance indices of the decision alternatives: The option with a utility rating of 100 is the best option (17). Other options are rated by best.

$$N_j = \frac{Q_j}{Q_{\max}} * 100\% \tag{17}$$

DEMATEL

The DEMATEL (The Decision-Making Trial and Evaluation Laboratory) method, which is a model analysis that reveals causal relationships with the help of matrices and to reach effect diagrams between a set of criteria. It defines the relations between the components with diagrams and matrices, makes quantitative definitions between these relations, and reveals the power between them (Bai & Sarkis, 2013).

Step 1. Obtaining the total relationship matrix. (S) is represented by the unit matrix (I) in Eq. (18). After subtracting the normalized direct relationship matrix from the unit matrix and inverting it, it is multiplied by itself again to obtain the total relationship matrix.

$$\begin{aligned} S &= M + M^2 + M^3 + \ldots = \sum\nolimits_{i=1}^{\leftarrow} M^i \\ &= M\,(\mathrm{I} - M)M^{-1} \end{aligned} \tag{18}$$

Step 2. Calculation of sender and receiver group. Sender and receiver groups are calculated (19, 20, and 21). The sum of the columns in the S matrix (R) represents the sum of the rows in the S matrix (D). By calculating the D and R equations and using the D-R and $D + R$ values, the effect of each criterion on the others and the level of relationship with the others are determined.

$$S = \begin{bmatrix} s_{11} & s_{1j} & s_{1n} \\ s_{i1} & s_{ij} & s_{in} \\ s_{n1} & s_{nj} & s_{nn} \end{bmatrix} \; i,j \in \{1, 2, 3, \ldots, n\} \tag{19}$$

$$D = \sum\nolimits_{j=1}^{n} S_{i,j} \tag{20}$$

$$R = \sum_{i=1}^{n} S_{i,j} \tag{21}$$

Step 3. Calculation of weights. With the help of $D + R$ and D-R values, weights are calculated by using Eqs. (22) and (23).

$$w_i = \left\{ (D_i + R_i)^2 + (D_i + R_i)^2 \right\}^{1/2} \tag{22}$$

$$w_i = \frac{w_i}{\sum_{i=1}^{n} w_i} \tag{23}$$

3 Applying the Proposed Framework

The Technology-Organization-Environment (TOE) framework is a theory that describes critical technological, organizational and environmental impacts on organizations' decisions (Tornatzky et al., 1990). It is based on describing behavioral considerations, interactions, and the application of innovation at the firm level. The technological dimension summarizes current technologies in use and new technologies relevant to each firm. While the organizational dimension reflects the characteristics of the firm in terms of scope, size, and resources, the environmental dimension indicates the environment in which business is conducted, including environmental impact, competitors, and government (Tornatzky et al., 1990). Some of the studies dealing with the TOE framework are as follows; Hashimy et al. (2023) described the adoption process for blockchain by borrowing multiple perspectives of technology, organization, and environment. Khan et al. (2021) observed the behavior of firms in adopting and using mobile payment systems with the support of the TOE framework. Peggy Ng et al. (2022) investigated the effectiveness of remote work practices and the factors affecting their results and evaluated them within the scope of TOE. Ullah et al. (2021) identified a multi-layered risk management framework for sustainable smart city management with the TOE framework. Dadhich and Hiran (2022) examined the impact of economic, social, technological, and organizational factors on corporate environmental sustainability. It is claimed to be the first known study to use TOE and TBL together (Dadhich & Hiran, 2022). The dimensions of the Triple Bottom Line (TBL) include the main economic, social, and environmental factors, which are the three important dimensions of sustainability (Elkington, 1997). In the current study, TBL dimensions and TOE framework were brought together to show the circular economy barriers in textile supply chain management (Fig. 2). Unlike the study of Dadhich

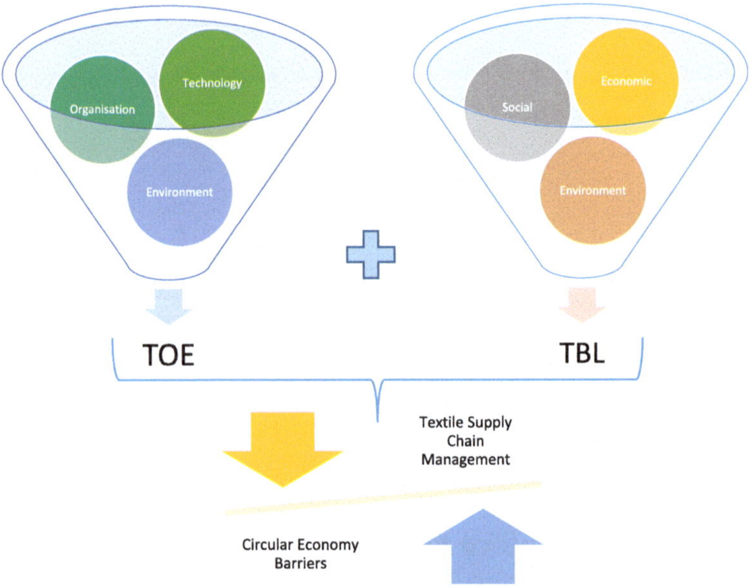

Fig. 2 TOE-TBL approach in textile supply chain circular economy barriers

and Hiran (2022), circular economy barriers are discussed in this perspective with more than one MCDM technique.

Determination of Criteria

Textile supply chain management is one of the processes in which the use of resources and, subsequently, the amount of waste is high. In this respect, it can be said that the textile industry needs the use of the circular economy, which takes resource management with the disciplines of "Reduce-Reuse-Recycle." However, although the concept has been noticed by researchers and company employees in recent years, it has brought some obstacles along with it, although it has increased the functioning of the circular economy. Addressing these obstacles with the TOE-TBL approach will reveal important parameters to be considered within the scope of the main dimension in the textile sector. In this sense, the first way to follow the conceptual model (Fig. 1) is a comprehensive literature review. The obstacles in front of the circular economy have been researched in the literature regardless of the sector, and then the circular economy studies of the textile sector have been focused. At the end of the criteria scanned in the literature, 786 total criteria were reached, and it decreased to 126 criteria by separating the criteria related to textile. Then repetitive and evocative criteria with inclusiveness were also removed, and 20 criteria were named as criteria from the literature. As a result of expert opinions, 2 more criteria were added, and a total of 22 criteria were included in the study (Table 1).

Table 1 Barriers to transition to circular economy model

TOE-TBL	Code	Circular economy barriers in textile	Authors
Economic	ECO1	Sustainability studies are not seen as an investment	Mishra et al. (2022)
	ECO2	Lack of funding for circular business models	Wuni (2022)
	ECO3	High cost of environmentally friendly products or materials	Gumley (2014), Wuni (2022)
	ECO4	Processing costs of cyclic products are higher than processed products	Rathinamoorthy (2019)
Environment	ENV1	Failure to implement effective take-back systems	Kalmykova et al. (2018)
	ENV2	Lack of environmental laws and regulations	Koszewska (2018), Vermunt et al. (2019), Wuni (2022)
	ENV3	Insufficient environmental education	Geng et al. (2009), Todeschini et al. (2017)
	ENV4	Uncertainty of the quality of recycled products	Koszewska (2018), Rathinamoorthy (2019), Bao and Lu (2020)
	ENV5	The challenge of recycling used textiles	Franco (2017), Koszewska (2018)
Organization	ORG1	Lack of strict compliance in circular economy practices	Singh et al. (2020)
	ORG2	Lack of transparency, communication, and information	Takacs et al. (2022), Vermunt et al. (2019), Wuni (2022), Koszewska (2018)
	ORG3	Lack of top management support and leadership	Manoharan et al. (2022), Agyemang et al. (2019), Singh et al. (2020)
	ORG4	Weak cooperation among stakeholders in the circular economy value chain	Wuni (2022)
	ORG5	Lack of qualified personnel and experts	Vermunt et al. (2019), Rathinamoorthy (2019), Govindan and Hasanagic (2018)
Social	SOC1	Difficulties in adapting to technological changes	Rizos and Bryhn (2022)
	SOC2	Uncertain and weak market demand for refurbished, remanufactured, and recycled products	Wuni (2022), Agyemang et al. (2019), Koszewska (2018)
	SOC3	Lack of awareness of those involved in the supply chain line	Kala and Bolia (2022), Dutta et al. (2021)
	SOC4	Resistance to change	Manoharan et al. (2022), Agyemang et al. (2019)
	SOC5	Existence of limited resource use in response to increasing population	Author
Technology	TEC1	Lack of technology to monitor product lifecycles	Guldmann and Huulgaard (2020)

(continued)

Table 1 (continued)

TOE-TBL	Code	Circular economy barriers in textile	Authors
	TEC2	Product remanufacturing and system design complexity	Majumdar and Sinha (2019)
	TEC3	Inadequate use of digital technologies	Author

Prioritization of Criteria

There are experts involved in the process of determining and collecting data in the study. Academicians and those working in the textile sector are included in the process. It was aimed that the fields of expertise of the academicians were supply chain management, sustainability, textile supply chain, and the knowledge of the circular economy concept was consulted. Thus, 12 academicians with different titles took part in the study. Persons with 10 years or more of experience in the field of processes for the textile industry were determined, and 6 people were reached. These people were asked to consider possible obstacles by giving general information about the concept of circular economy. Focus group meetings with an expert group of 18 people were held in the form of remote interviews. After the final circular economy barrier table was created, it was started to analyze with MCDM techniques. First of all, for the SWARA technique, it was requested to rank the criteria according to the degree of importance between 1 and 22. Scoring between 1 and 4 for the DEMATEL technique was requested, and the comparison of the criteria was requested and evaluated in Excel format. For COPRAS and TOPSIS, it has been determined by scoring from 1 to 9 within the criteria of technological maturity, cost awareness, sustainability, and social acceptability within the framework of TOE-TBL. For AHP, Saaty's importance rating scale was used and the criteria between 1 and 9 were asked to score, and the incoming evaluations were used in all three methods and the findings were shown (Tables 2, 3, and 5).

The purpose of applying different MCDM methods in the study was to observe whether possible criteria would express similar results. MCDM techniques have different consistencies within themselves, and it is aimed to reveal similarities and differences by using different MCDM techniques. Additionally, an overall average was taken for TOE-TBL and an evaluation was made.

In line with the decision matrices, the steps of the MCDM techniques were applied and the importance levels of the circular economy barriers were listed for each method (Table 4). According to the results of the SWARA method, the most important obstacle was "Lack of environmental laws and regulations." According to the COPRAS and AHP method, the "Uncertain and weak market demand for refurbished, remanufactured and recycled products" barrier was determined as the most important result. According to the TOPSIS and DEMATEL methods, the most important obstacle was determined as "Uncertainty of the quality of recycled products." The barrier of "Existence of limited resource use in response to increasing population" is at the bottom of the list for TOPSIS, COPRAS, AHP, and

Table 2 TOPSIS, AHP, SWARA priority values

	TOPSIS			AHP	SWARA
Code	S_i*	S_i-	C_i*	W_a	W_s
ECO1	0.0066	0.1415	0.0443	0.013	0.01
ECO2	0.0395	0.1165	0.2533	0.058	0.06
ECO3	0.0272	0.1317	0.1714	0.044	0.03
ECO4	0.0087	0.1377	0.0591	0.015	0.02
ENV1	0.0055	0.1408	0.0374	0.015	0.01
ENV2	0.0677	0.0857	0.4413	0.087	0.11
ENV3	0.0313	0.1269	0.1977	0.040	0.03
ENV4	0.1209	0.0443	0.7317	0.104	0.10
ENV5	0.0485	0.1135	0.2996	0.053	0.04
ORG1	0.0484	0.1123	0.3013	0.062	0.10
ORG2	0.0108	0.1384	0.0721	0.029	0.03
ORG3	0.0333	0.1242	0.2112	0.051	0.07
ORG4	0.0123	0.1363	0.0827	0.024	0.02
ORG5	0.0257	0.1287	0.1663	0.045	0.05
SOC1	0.0664	0.0971	0.4064	0.067	0.08
SOC2	0.1096	0.0563	0.6606	0.122	0.09
SOC3	0.0637	0.1025	0.3833	0.057	0.04
SOC4	0.0138	0.1346	0.0931	0.022	0.02
SOC5	0.0018	0.1433	0.0123	0.010	0.01
TEC1	0.0106	0.1382	0.0711	0.026	0.02
TEC2	0.0288	0.1269	0.1848	0.046	0.04
TEC3	0.0056	0.1411	0.0379	0.013	0.02

DEMATEL. Only according to the SWARA method, the "Sustainability studies are not seen as an investment" obstacle is in the last place.

According to the results of TOPSIS, COPRAS, AHP, SWARA, and DEMATEL methods, the top five circular economy barriers in textile supply chain management are "Lack of environmental laws and regulations," "Uncertainty of the quality of recycled products," "Lack of strict compliance in circular economy practices," "Difficulties in adapting to technological changes," "Uncertain and weak market demand for refurbished, remanufactured and recycled products," and "Lack of awareness of those involved in the supply chain line" criteria are among the most important obstacles (Table 4). When Fig. 3 is examined, it can be interpreted that similar criteria are in a similar order when attention is paid to the color distinctions and transitions between the obstacles. Circular economy barriers containing a harmonic order will be evaluated with similar priority within MCDM techniques and will give an idea to textile stakeholders.

Interactions of Criteria

In addition to ranking the circular economy barriers, which are considered from the perspective of textile supply chain management, according to their weights, the effect levels between the barriers were investigated with the help of the DEMATEL

Table 3 COPRAS priority values display

Code	$S+_i$	$S-_i$	$S-_{min}$	Total S_{-i}	$S-_{min}/S-_i$	Total $S-_{min}/S-_i$	Q_i	P_i
ECO1	0.000	0.0143	0.0112	1.0000	0.7807	8.7744	0.0890	78.07
ECO2	0.000	0.0532			0.2104		0.0240	21.04
ECO3	0.000	0.0354			0.3161		0.0360	31.61
ECO4	0.000	0.0176			0.6365		0.0725	63.65
ENV1	0.000	0.0147			0.7614		0.0868	76.14
ENV2	0.000	0.0864			0.1296		0.0148	12.96
ENV3	0.000	0.0374			0.2991		0.0341	29.91
ENV4	0.000	0.1268			0.0883		0.0101	8.83
ENV5	0.000	0.0533			0.2100		0.0239	21.00
ORG1	0.000	0.0574			0.1950		0.0222	19.50
ORG2	0.000	0.0214			0.5226		0.0596	52.26
ORG3	0.000	0.0443			0.2525		0.0288	25.25
ORG4	0.000	0.0217			0.5160		0.0588	51.60
ORG5	0.000	0.0365			0.3069		0.0350	30.69
SOC1	0.000	0.0736			0.1521		0.0173	15.21
SOC2	0.000	0.1339			0.0836		0.0095	8.36
SOC3	0.000	0.0656			0.1706		0.0194	17.06
SOC4	0.000	0.0219			0.5113		0.0583	51.13
SOC5	0.000	0.0112			1.0000		0.1140	100.00
TEC1	0.000	0.0207			0.5408		0.0616	54.08
TEC2	0.000	0.0384			0.2915		0.0332	29.15
TEC3	0.000	0.0140			0.7994		0.0911	79.94

method (Table 5). In this context, the following barriers have been identified as the affected barriers:

- ECO1 Sustainability studies are not seen as an investment
- ECO4 Processing costs of cyclic products are higher than processed products
- ENV1 Failure to implement effective take-back systems
- ENV3 Insufficient environmental education
- ENV4 Uncertainty of the quality of recycled products
- ORG1 Lack of strict compliance in circular economy practices
- SOC2 Uncertain and weak market demand for refurbished, remanufactured and recycled products
- SOC4 Resistance to change
- SOC5 Existence of limited resource use in response to increasing population
- TEC2 Product remanufacturing and system design complexity

The barriers affecting these barriers are as follows:

- ECO2 Lack of funding for circular business models
- ECO3 High cost of environmentally friendly products or materials
- ENV2 Lack of environmental laws and regulations
- ENV5 The challenge of recycling used textiles

Table 4 Sequencing of barriers with MCDM techniques

Code	Circular economy barriers in textile	1	2	3	4	5
ECO1	Sustainability studies are not seen as an investment	19	20	20	22	19
ECO2	Lack of funding for circular business models	8	8	6	7	6
ECO3	High cost of environmentally friendly products or materials	12	13	12	12	15
ECO4	Processing costs of cyclic products are higher than processed products	18	18	18	19	18
ENV1	Failure to implement effective take-back systems	21	19	19	21	20
ENV2	Lack of environmental laws and regulations	3	3	3	1	2
ENV3	Insufficient environmental education	10	11	13	14	14
ENV4	Uncertainty of the quality of recycled products	1	2	2	2	1
ENV5	The challenge of recycling used textiles	7	7	8	11	8
ORG1	Lack of strict compliance in circular economy practices	6	6	5	3	3
ORG2	Lack of transparency, communication, and information	16	16	14	13	9
ORG3	Lack of top management support and leadership	9	9	9	6	10
ORG4	Weak cooperation among stakeholders in the circular economy value chain	15	15	16	17	11
ORG5	Lack of qualified personnel and experts	13	12	10	8	13
SOC1	Difficulties in adapting to technological changes	4	4	4	5	4
SOC2	Uncertain and weak market demand for refurbished, remanufactured, and recycled products	2	1	1	4	5
SOC3	Lack of awareness of those involved in the supply chain line	5	5	7	9	7
SOC4	Resistance to change	14	14	17	16	12
SOC5	Existence of limited resource use in response to increasing population	22	22	22	20	22
TEC1	Lack of technology to monitor product lifecycles	17	17	15	15	16
TEC2	Product remanufacturing and system design complexity	11	10	11	10	17
TEC3	Inadequate use of digital technologies	20	21	21	18	21

(*1: TOPSIS 2: COPRAS 3: AHP 4: SWARA 5: DEMATEL)

- ORG2 Lack of transparency, communication and information
- ORG3 Lack of top management support and leadership
- ORG4 Weak cooperation among stakeholders in the circular economy value chain
- ORG5 Lack of qualified personnel and experts
- SOC1 Difficulties in adapting to technological changes
- SOC3 Lack of awareness of those involved in the supply chain line
- TEC1 Lack of technology to monitor product lifecycles
- TEC3 Inadequate use of digital technologies

If we explain the interactions with examples in the direction of the results. The lack of financing of circular business models affects the fact that sustainability studies are not seen as an investment. Insufficient financial resources also hinder investments and affect sustainability, that is, circular economy studies. Similarly, the lack of environmental laws and regulations affects the lack of effective take-back systems, the uncertain quality of recycled products, and the lack of strict compliance

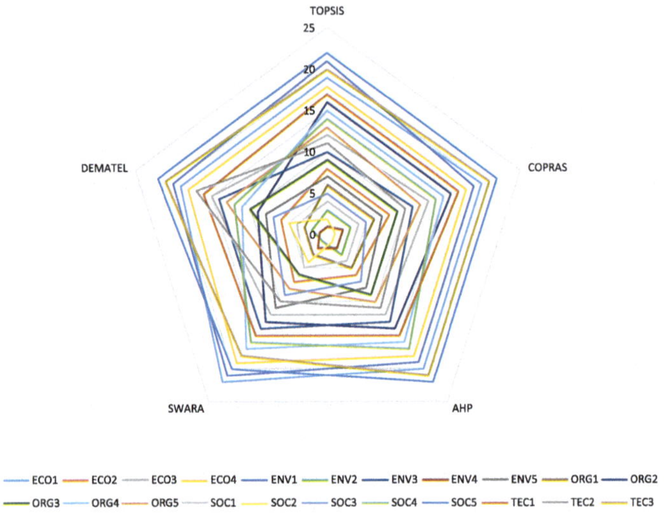

Fig. 3 Display of rank change of barriers

in circular economy practices. The increase in the environmental studies of the laws standardizes the returns from the products and ensures that they are recycled effectively. Again, it can eliminate the poor quality understanding toward customers and suppliers. Strict compliance with laws can contribute to the transformation of the circular economy. On the other hand, the lack of top management support and leadership and the lack of information and communication affect the resistance to change. As can be seen in Fig. 4, the distribution of the affected and affecting barriers shows the ones in the positive section, while those in the negative section represent the affected ones.

4 The Place of Barriers in the TOE-TBL Approach

While the TOE framework, which was put forward by considering the textile sector, defines the critical technological, organizational, and environmental effects on the decisions of the organizations, the TBL approach deals with economic, social, and environmental dimensions and circular economy barriers. Total 22 criteria were classified according to TOE-TBL dimensions and ranked and evaluated with MCDM techniques. According to this ranking (Table 4), the averages of the rankings

Table 5 DEMATEL priority values and impact values

Code	D_i	R_j	$D_i + R_j$	$D_i–R_j$	Impact group
ECO1	2.31372	3.75338	6.0671	−1.4397	Affected
ECO2	4.13015	2.64058	6.77073	1.48957	Affecting
ECO3	3.11417	3.0338	6.14797	0.08037	Affecting
ECO4	2.84174	3.27153	6.11327	−0.4298	Affected
ENV1	2.7408	3.31831	6.05911	−0.5775	Affected
ENV2	3.68403	3.31345	6.99748	0.37058	Affecting
ENV3	2.91726	3.25044	6.16769	−0.3332	Affected
ENV4	3.33937	3.72623	7.0656	−0.3869	Affected
ENV5	3.4322	3.19515	6.62735	0.23705	Affecting
ORG1	3.14305	3.71392	6.85697	−0.5709	Affected
ORG2	3.29865	3.18763	6.48628	0.11103	Affecting
ORG3	3.24594	3.21956	6.4655	0.02638	Affecting
ORG4	3.22736	3.10064	6.328	0.12673	Affecting
ORG5	3.24482	2.97789	6.22271	0.26693	Affecting
SOC1	3.90386	2.91768	6.82155	0.98618	Affecting
SOC2	3.28687	3.50889	6.79576	−0.222	Affected
SOC3	3.60367	3.0669	6.67057	0.53677	Affecting
SOC4	2.92342	3.36516	6.28858	−0.4417	Affected
SOC5	2.72638	2.7591	5.48548	−0.0327	Affected
TEC1	3.26327	2.87058	6.13385	0.39269	Affecting
TEC2	2.78313	3.34873	6.13186	−0.5656	Affected
TEC3	3.122	2.74631	5.86831	0.37569	Affecting

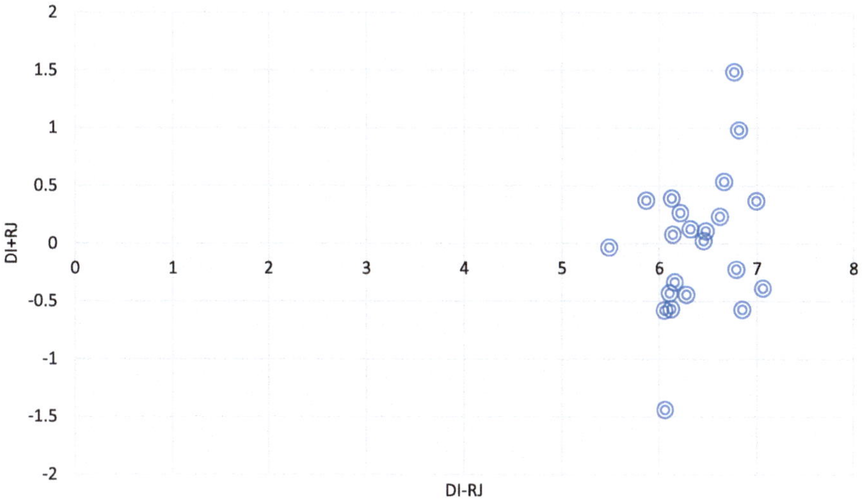

Fig. 4 Impact display of circular economy model barriers

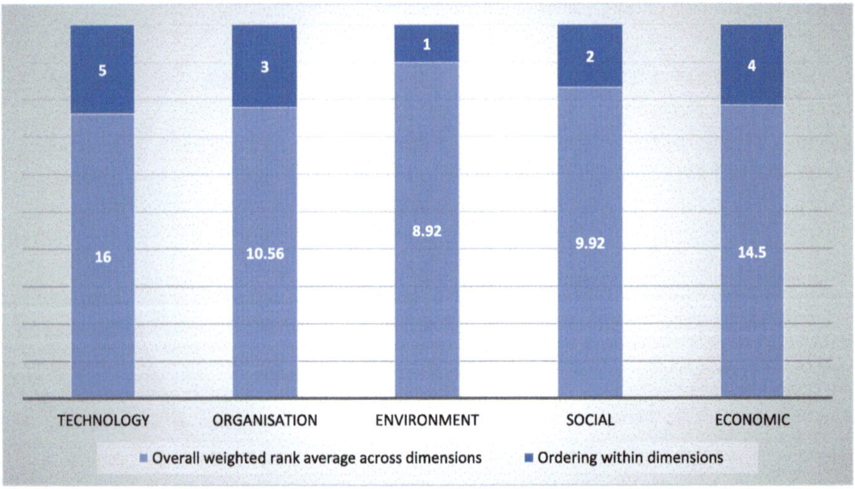

Fig. 5 Ranking of importance levels for TOE-TBL

were taken as the Technology group within itself, the Organization group within itself, and the other groups within itself, and the companies were asked to decide from which main dimension they would start in their prioritization studies (Fig. 5). The environmental dimensions of the results were in the first place as the criteria that companies that will transition to the circular economy model should prioritize. Following this, the social factor is in the second place, the organization factor is in the third place, the economic factor is in the fourth place, and finally, the technology factor is in the last place.

5 Theoretical and Practical Implication

Since the environmental factor is included in both TOE and TBL dimensions, it is a significant factor with a high impact level. Environmental factors may be the area where such obstacles, which play an important role in the transition to the circular economy model, can be approached with a solution-oriented approach. The inability to implement effective take-back systems, which are included in environmental factors, the absence of environmental laws and regulations, insufficient environmental education, the uncertainty of the quality of recycled products, and the difficulty of recycling used textiles are an important part of the circular economy issue where resource management is important. With the circular economy, which sets out with the principles of reduce, reuse, and recycle, and if the barriers of environmental factors are reduced or eliminated, the companies in the sector will be able to increase productivity through different studies such as resource recycling, quality reliability, and the creation of effective take-back systems. Another of the studies that companies have been dealing with recently is corporate sustainability studies, and the focus may be on social factors. Areas open to improvement in this study are the difficulties experienced in adapting to technological changes, which are among the social factors, uncertain and weak market demand for renewed, remanufactured and recycled products, lack of awareness of those involved in the supply chain line, resistance to change, limited resource use in response to the increasing population have affected the textile industry. Possible breakthroughs toward society, stakeholders, and employees will be able to raise social factors to higher levels. With the increase in the quality of recycled products, the society's demand for products with circular economy will increase, and the level of awareness will increase with the dissemination of necessary trainings. On the other hand, balancing the population growth will be ensured by family planning and resource consumption will be limited by preventing excessive population growth. The criteria included in the main factor of the organization targeting organizational culture are listed as lack of strict compliance in circular economy practices; lack of transparency, communication, and information; lack of senior management support and leadership; weak cooperation between stakeholders in the circular economy value chain; and lack of qualified personnel and experts. The most important issues seem to be the lack of strict compliance and the lack of support from the top management, and if circular economy models are adopted as a solution proposal and standards are provided throughout the enterprise, significant gains will be achieved. Thus, resource management will be easier and more efficient with the increasing importance of communication and information among stakeholders. However, with the reorganization of environmental laws, studies can be carried out to eliminate possible compliance problems and employees will be able to improve themselves in this regard. Sustainability studies, which are among the economic factors, are not seen as an investment; lack of financing for circular business models, high cost of environmentally friendly products or materials, processing costs of circular products, and being higher than processed products cause the sector to look at it as an obstacle to its development in this area. Making the circular economy feel not a choice but a necessity will bring investments and cost planning to the agenda again. Last, but not least, another major hurdle as important as the first factor is technology and includes criteria such as lack of technology to monitor product lifecycles, product remanufacturing and system design complexity, and inadequate use of digital technologies. Acquiring software and technological devices that will provide possible improvements without being indifferent to the developing technology will be an important step toward the future. Technologies that will contribute to the life cycle can be preferred, since we can take mitigation actions if we know what the environmental impacts of the products are, and following up-to-date technologies such as artificial intelligence applications, IoT, and Blockchain technologies that facilitate traceability will transform many parameters into positive ones.

6 Conclusions and Discussions

The textile industry is one of the largest industries worldwide and is responsible for generating large amounts of waste and environmental impact. Therefore, applying circular economy principles in the textile industry can provide both environmental and economic benefits. However, reaching this economy model brings with it many obstacles and slows down the circular economy studies. This study aimed to determine the parameters that will enable us to take action by investigating the circular economy barriers with a comprehensive study. The evaluation table, consisting of 5 dimensions and 22 sub-obstacles within the TOE-TBL framework, was evaluated with MCDM techniques, and the lack of environmental laws and regulations, the uncertainty of the quality of recycled products, and the uncertain and weak market demand for renewed, remanufactured, and recycled products were the first areas of action to be addressed. Has emerged. According to the TOE-TBL approach, environmental studies have come to the fore in the main dimension of the obstacles to the circular economy.

While economic, social, environmental, technological, and organization are the main classification areas in the current study, Rathinamoorthy (2019) addressed the circular economy barriers in the textile and apparel industry as cultural, technological, market, and government regulations. Kazancoğlu et al. (2020) distinguished the circular economy barriers of the textile supply chain as management and decision-making, labor, design challenges, materials, rules and regulations, lack of knowledge and awareness, lack of integration and cooperation, and cost and technical infrastructure. According to Saccani et al. (2023), while making this distinction, classified circular economy adoption barriers in Italian textile and fashion supply chains as economic and financial viability, market and competition, product features, standards and regulation, supply chain management, and user behavior, Bressanelli et al. (2019) benefited from. In addition, Jia et al. (2020), in their literature review, separated the obstacles of the textile and ready-made clothing industry as Organizational, Financial, and Policy and stated that the most important obstacle is the Lack of financial support. Dulia et al. (2021) classified circular supply chain risks in the Bangladesh textile industry as government, economic, technological, knowledge and skills, organizational, social and cultural, market, financial, and, finally logistics. Decreased quality of recycled products, lack of a proper regulatory vision of the circular economy and not considering the needs of the entire textile supply chain were found to be high-priority barriers. In the current study, the uncertainty of the quality of the recycled products took the first place, strengthening the accuracy of possible barriers. Huang et al. (2021) applied a similar classification to the current study, dividing the barriers to the circular economy in Taiwan's textile sector into official and regulatory, economic and financial, technological, social, organizational and managerial, infrastructure, and supply chain and market. However, in the current study, there was a differentiation from this study by making the TOE-TBL distinction. In the study, low customer demand for recycled textiles, lack of successful circular business models, difficulty of collaborative innovation among supply chain partners, lack of high-quality recycling materials, and high costs but low economic benefits in the short term were identified as primary barriers. In the current study, quality uncertainty, lack of market demand, lack of circular business models, and lack of financing were determined as priority obstacles, and its accuracy was ensured. Again, Majumdar et al. (2022) identified 10 causal barriers using Grey-DEMATEL analysis of major circular barriers in the textile and apparel supply chain. Lack of supply chain collaboration, lack of effective planning for the circular supply chain, lack of incentives for circularity in the supply chain, and lack of technical knowledge for recycling have emerged as major causal barriers. In the current study, the lack of awareness was in the top five in the ranking. Hartley et al. (2022) classified the barriers to the circular economy in the Dutch textile industry according to cultural, regulatory, market, and technical domains. The high costs of producing and marketing circular products, the lack of awareness and interest of consumers, the still dominant linear mindset of companies operating in the value chain, and regulations and laws were identified as primary barriers. In the current study, lack of awareness, lack of strict compliance, and lack of environmental laws and regulations were identified as important obstacles and its accuracy was ensured. Franco (2017) explored the circular economy challenges in the textile and apparel industry, and the technological challenges of recycling post-consumer textiles and poor working standards were the leading hurdles. In the current study, the difficulties experienced in adapting to technological changes took place in the top five and increased the level of importance once again.

As a result of the studies carried out, the literature-supported results and the environmental factors being the first study area to overcome possible obstacles according to the TOE-TBL framework provide valuable contributions to those who will benefit from the study. The results obtained guide the textile supply chain management stakeholders. On the other hand, it has the quality to provide information exchange even to decision makers, policy makers, and organizations operating in the textile field. The textile industry has great potential for circular economy practices, and the adoption of these principles can reduce the industry's environmental impact and increase sustainability.

The present study also has some limitations. First of all, all the participants in the study made comments and inferences on the textile sector, taking into account the generality of Turkey. The results may be interpreted differently in more developed countries. Disseminating these results when the findings are discussed may reduce the accuracy of these results, but when compared with the results in different regions in the discussion texts, reliability is increased with similar results. Another issue may be that the difficulties encountered in the transition to the circular economy model, as the study targets the textile sector, are related to this field and not to other sectors. Finally, as the barriers discussed cover the literature and expert opinion, they may include some biased behavior and overlooked barriers. In this way, future studies with different obstacles and different working methods can close these gaps.

Ethical Approval and Consent to Participate Not applicable.

Competing Interests The author declares no competing interests.

Author Contributions EY: Conceptualization, methodology, validation, writing–original draft preparation, writing–reviewing and editing.

Consent to Publish Not applicable.

Funding The authors declare that no funds, grants, or other support were received during the preparation of this manuscript.

Data Availability Availability of data, the data that support the findings of this study are available from the corresponding author.

References

Abdelmeguid, A., Afy-Shararah, M., & Salonitis, K. (2022). Investigating the challenges of applying the principles of the circular economy in the fashion industry: A systematic review. *Sustainable Production and Consumption, 32*, 505–518.

Abishini, A. H., & Karthikeyan, K. M. B. (2023). Application of MCDM and Taguchi super ranking concept for materials selection problem. *Materials Today: Proceedings, 72*, 2480–2487.

Agyemang, M., Kusi-Sarpong, S., Khan, S. A., Mani, V., Rehman, S. T., & Kusi-Sarpong, H. (2019). *Drivers and barriers to circular economy implementation: An explorative study in Pakistan's automobile industry* (Vol. 57, p. 971). Management Decision.

Akter, M. M. K., Haq, U. N., Islam, M. M., & Uddin, M. A. (2022). Textile-apparel manufacturing and material waste management in the circular economy: A conceptual model to achieve sustainable development goal (SDG) 12 for Bangladesh. *Cleaner Environmental Systems, 4*, 100070.

Alimardani, M., Hashemkhani Zolfani, S., Aghdaie, M. H., & Tamošaitienė, J. (2013). A novel hybrid SWARA and VIKOR methodology for supplier selection in an agile environment. *Technological and Economic Development of Economy, 19*(3), 533–548.

Al-Tohamy, R., Ali, S. S., Li, F., Okasha, K. M., Mahmoud, Y. A. G., Elsamahy, T., et al. (2022). A critical review on the treatment of dye-containing wastewater: Ecotoxicological and health concerns of textile dyes and possible remediation approaches for environmental safety. *Ecotoxicology and Environmental Safety, 231*, 113160.

Ayçin, E. (2020). *Çok kriterli karar verme: Bilgisayar uygulamalı çözümler*. Nobel Akademik Yayıncılık.

Bai, C., & Sarkis, J. (2013). A grey-based DEMATEL model for evaluating business process management critical success factors. *International Journal of Production Economics, 146*(1), 281–292.

Balanay, R., & Halog, A. (2019). Tools for circular economy: Review and some potential applications for the Philippine textile industry. In *Circular economy in textiles and apparel* (pp. 49–75).

Bao, Z., & Lu, W. (2020). Developing efficient circularity for construction and demolition waste management in fast emerging economies: Lessons learned from Shenzhen, China. *Science of the Total Environment, 724*, 138264.

Bressanelli, G., Perona, M., & Saccani, N. (2019). Challenges in supply chain redesign for the circular economy: A literature review and a multiple case study. *International Journal of Production Research, 57*(23), 7395–7422.

Chen, L., Wang, L., Wu, X., & Ding, X. (2017). A process-level water conservation and pollution control performance evaluation tool of cleaner production technology in textile industry. *Journal of Cleaner Production, 143*, 1137–1143.

Dadhich, M., & Hiran, K. K. (2022). Empirical investigation of extended TOE model on corporate environment sustainability and dimensions of operating performance of SMEs: A high order PLS-ANN approach. *Journal of Cleaner Production, 363*, 132309.

Dulia, E. F., Ali, S. M., Garshasbi, M., & Kabir, G. (2021). Admitting risks towards circular economy practices and strategies: An empirical test from supply chain perspective. *Journal of Cleaner Production, 317*, 128420.

Dutta, P., Talaulikar, S., Xavier, V., & Kapoor, S. (2021). Fostering reverse logistics in India by prominent barrier identification and strategy implementation to promote circular economy. *Journal of Cleaner Production, 294*, 126241.

Elkington, J. (1997). The triple bottom line. *Environmental management: Readings and cases, 2*, 49–66.

Ellen MacArthur Foundation. (2017). *A new textiles economy: Redesigning fashion's future*. http://www.ellenmacarthurfoundation.org/publications

Filho Leal, W., Ellams, D., Han, S., Tyler, D., Boiten, V. J., Paço, A., et al. (2019). A review of the socio-economic advantages of textile recycling. *Journal of Cleaner Production, 218*, 10–20.

Fischer, A., & Pascucci, S. (2017). Institutional incentives in circular economy transition: The case of material use in the Dutch textile industry. *Journal of Cleaner Production, 155*, 17–32.

Franco, M. A. (2017). Circular economy at the micro level: A dynamic view of incumbents' struggles and challenges in the textile industry. *Journal of Cleaner Production, 168*, 833–845.

Galatti, L. G., & Baruque-Ramos, J. (2022). Circular economy indicators for measuring social innovation in the Brazilian textile and fashion industry. *Journal of Cleaner Production, 363*, 132485.

Geng, Y., Zhu, Q., Doberstein, B., & Fujita, T. (2009). Implementing China's circular economy concept at the regional level: A review of progress in Dalian, China. *Waste Management, 29*(2), 996–1002.

Gong, Z. T., & Shi, Z. H. (2008, July). The TOPSIS method based on covering rough sets. In *2008 international conference on machine learning and cybernetics* (Vol. 4, pp. 2430–2433). IEEE.

Govindan, K., & Hasanagic, M. (2018). A systematic review on drivers, barriers, and practices towards circular economy: A supply chain perspective. *International Journal of Production Research, 56*(1–2), 278–311.

Guldmann, E., & Huulgaard, R. D. (2020). Barriers to circular business model innovation: A multiple-case study. *Journal of Cleaner Production, 243*, 118160.

Gumley, W. (2014). An analysis of regulatory strategies for recycling and re-use of metals in Australia. *Resources, 3*(2), 395–415.

Hartley, K., Roosendaal, J., & Kirchherr, J. (2022). Barriers to the circular economy: The case of the Dutch technical and interior textiles industries. *Journal of Industrial Ecology, 26*(2), 477–490.

Hashimy, L., Jain, G., & Grifell-Tatjé, E. (2023). Determinants of blockchain adoption as decentralized business model by Spanish firms–an innovation theory perspective. *Industrial Management & Data Systems, 123*(1), 204–228.

Huang, Y. F., Azevedo, S. G., Lin, T. J., Cheng, C. S., & Lin, C. T. (2021). Exploring the decisive barriers to achieve circular economy: Strategies for the textile innovation in Taiwan. *Sustainable Production and Consumption, 27*, 1406–1423.

Jäämaa, L., & Kaipia, R. (2022). The first mile problem in the circular economy supply chains–collecting recyclable textiles from consumers. *Waste Management, 141*, 173–182.

Jia, F., Yin, S., Chen, L., & Chen, X. (2020). The circular economy in the textile and apparel industry: A systematic literature review. *Journal of Cleaner Production, 259*, 120728.

John, E. P., & Mishra, U. (2023). A sustainable three-layer circular economic model with controllable waste, emission, and wastewater from the textile and fashion industry. *Journal of Cleaner Production, 388*, 135642.
Kala, K., & Bolia, N. B. (2022). Analysis of informal waste management using system dynamic modelling. *Heliyon, 8*(8), e09993.
Kalmykova, Y., Sadagopan, M., & Rosado, L. (2018). Circular economy–from review of theories and practices to development of implementation tools. *Resources, Conservation and Recycling, 135*, 190–201.
Kayikci, Y., Kazancoglu, Y., Gozacan-Chase, N., Lafci, C., & Batista, L. (2022). Assessing smart circular supply chain readiness and maturity level of small and medium-sized enterprises. *Journal of Business Research, 149*, 375–392.
Kazancoglu, I., Kazancoglu, Y., Yarimoglu, E., & Kahraman, A. (2020). A conceptual framework for barriers of circular supply chains for sustainability in the textile industry. *Sustainable Development, 28*(5), 1477–1492.
Keršuliene, V., Zavadskas, E. K., & Turskis, Z. (2010). Selection of rational dispute resolution method by applying new step-wise weight assessment ratio analysis (SWARA). *Journal of Business Economics and Management, 11*(2), 243–258.
Khan, N. A., Khan, A. N., Bahadur, W., & Ali, M. (2021). Mobile payment adoption: A multi-theory model, multi-method approach and multi-country study. *International Journal of Mobile Communications, 19*(4), 467–491.
Koszewska, M. (2018). Circular economy—Challenges for the textile and clothing industry. *Autex Research Journal, 18*(4), 337–347.
Kumar, V., Sezersan, I., Garza-Reyes, J. A., Gonzalez, E. D., & Al-Shboul, M. D. A. (2019). Circular economy in the manufacturing sector: Benefits, opportunities and barriers. *Management Decision, 57*(4), 1067–1086.
Luoma, P., Penttinen, E., Tapio, P., & Toppinen, A. (2022). Future images of data in circular economy for textiles. *Technological Forecasting and Social Change, 182*, 121859.
Majumdar, A., & Sinha, S. K. (2019). Analyzing the barriers of green textile supply chain management in Southeast Asia using interpretive structural modeling. *Sustainable Production and Consumption, 17*, 176–187.
Majumdar, A., Ali, S. M., Agrawal, R., & Srivastava, S. (2022). A triple helix framework for strategy development in circular textile and clothing supply chain: An Indian perspective. *Journal of Cleaner Production, 367*, 132954.
Manoharan, S., Pulimi, V. S. K., Kabir, G., & Ali, S. M. (2022). Contextual relationships among drivers and barriers to circular economy: An integrated ISM and DEMATEL approach. *Sustainable Operations and Computers, 3*, 43–53.
Mishra, R., Singh, R. K., & Govindan, K. (2022). Barriers to the adoption of circular economy practices in micro, small and medium enterprises: Instrument development, measurement and validation. *Journal of Cleaner Production, 351*, 131389.
Neto de Oliveira, G. C., Correia, J. M. F., Tucci, H. N. P., Librantz, A. F. H., Giannetti, B. F., & de Almeida, C. M. V. B. (2022). Sustainable resilience degree assessment of the textile industrial by size: Incremental change in cleaner production practices considering circular economy. *Journal of Cleaner Production, 380*, 134633.
Ozkan-Ozen, Y. D., Kazancoglu, Y., & Mangla, S. K. (2020). Synchronized barriers for circular supply chains in industry 3.5/industry 4.0 transition for sustainable resource management. *Resources, Conservation and Recycling, 161*, 104986.
Palacios-Mateo, C., van der Meer, Y., & Seide, G. (2021). Analysis of the polyester clothing value chain to identify key intervention points for sustainability. *Environmental Sciences Europe, 33*(1), 2.
Peggy Ng, M., Lit, K. K., & Cheung, C. T. (2022). Remote work as a new normal? The technology-organization-environment (TOE) context. *Technology in Society, 70*, 102022.

Poursoltan, L., Mohammad Seyedhosseini, S., & Jabbarzadeh, A. (2021). A two-level closed-loop supply chain under the constract of vendor managed inventory with learning: A novel hybrid algorithm. *Journal of Industrial and Production Engineering, 38*(4), 254–270.

Quintás, M. A., Martínez-Senra, A. I., & Sartal, A. (2018). The role of SMEs' green business models in the transition to a low-carbon economy: Differences in their design and degree of adoption stemming from business size. *Sustainability, 10*(6), 2109.

Ranta, V., Aarikka-Stenroos, L., Ritala, P., & Mäkinen, S. J. (2018). Exploring institutional drivers and barriers of the circular economy: A cross-regional comparison of China, the US, and Europe. *Resources, Conservation and Recycling, 135*, 70–82.

Rathinamoorthy, R. (2019). Circular fashion. In *Circular economy in textiles and apparel* (pp. 13–48). Woodhead Publishing.

Rizos, V., & Bryhn, J. (2022). Implementation of circular economy approaches in the electrical and electronic equipment (EEE) sector: Barriers, enablers and policy insights. *Journal of Cleaner Production, 338*, 130617.

Ruzgys, A., Volvačiovas, R., Ignatavičius, Č., & Turskis, Z. (2014). Integrated evaluation of external wall insulation in residential buildings using SWARA-TODIM MCDM method. *Journal of Civil Engineering and Management, 20*(1), 103–110.

Saaty, T. L. (1988). *What is the analytic hierarchy process?* (pp. 109–121). Springer.

Saccani, N., Bressanelli, G., & Visintin, F. (2023). Circular supply chain orchestration to overcome circular economy challenges: An empirical investigation in the textile and fashion industries. *Sustainable Production and Consumption, 35*, 469–482.

Salmi, A., & Kaipia, R. (2022). Implementing circular business models in the textile and clothing industry. *Journal of Cleaner Production, 378*, 134492.

Shirvanimoghaddam, K., Motamed, B., Ramakrishna, S., & Naebe, M. (2020). Death by waste: Fashion and textile circular economy case. *Science of the Total Environment, 718*, 137317.

Silva, W. D. O., & Morais, D. C. (2022). Impacts and insights of circular business models' outsourcing decisions on textile and fashion waste management: A multi-criteria decision model for sorting circular strategies. *Journal of Cleaner Production, 370*, 133551.

Singh, R. K., Kumar, A., Garza-Reyes, J. A., & de Sá, M. M. (2020). Managing operations for circular economy in the mining sector: An analysis of barriers intensity. *Resources Policy, 69*, 101752.

Sadowski, A., Dobrowolska, B., Skowron-Grabowska, B., & Bujak, A. (2021). Polish textile and apparel industry: Global supply chain management perspective. *Autex Research Journal, 21*(3), 262–271.

Stanujkic, D., Karabasevic, D., & Zavadskas, E. K. (2015). A framework for the selection of a packaging design based on the SWARA method. *Engineering Economics, 26*(2), 181–187.

Takacs, F., Brunner, D., & Frankenberger, K. (2022). Barriers to a circular economy in small-and medium-sized enterprises and their integration in a sustainable strategic management framework. *Journal of Cleaner Production, 362*, 132227.

Teixeira, T. G. B., de Medeiros, J. F., Kolling, C., Ribeiro, J. L. D., & Morea, D. (2023). Redesign in the textile industry: Proposal of a methodology for the insertion of circular thinking in product development processes. *Journal of Cleaner Production, 397*, 136588.

Tektas, A., & Alakavuk, E. D. (2003). Allocation model: A tool to develop effective media plans for Turkey. *International Journal of Advertising, 22*(3), 333–348.

Todeschini, B. V., Cortimiglia, M. N., Callegaro-de-Menezes, D., & Ghezzi, A. (2017). Innovative and sustainable business models in the fashion industry: Entrepreneurial drivers, opportunities, and challenges. *Business Horizons, 60*(6), 759–770.

Tornatzky, L. G., Fleischer, M., & Chakrabarti, A. K. (1990). *Processes of technological innovation*. Lexington Books.

Ullah, F., Qayyum, S., Thaheem, M. J., Al-Turjman, F., & Sepasgozar, S. M. (2021). Risk management in sustainable smart cities governance: A TOE framework. *Technological Forecasting and Social Change, 167*, 120743.

Vermunt, D. A., Negro, S. O., Verweij, P. A., Kuppens, D. V., & Hekkert, M. P. (2019). Exploring barriers to implementing different circular business models. *Journal of Cleaner Production, 222*, 891–902.

Villacreses, G., Martinez-Gomez, J., Jijon, D., & Cordovez, M. (2022). Geolocation of photovoltaic farms using geographic information systems (GIS) with multiple-criteria decision-making (MCDM) methods: Case of the Ecuadorian energy regulation. *Energy Reports, 8*, 3526–3548.

Wuni, I. Y. (2022). Mapping the barriers to circular economy adoption in the construction industry: A systematic review, Pareto analysis, and mitigation strategy map. *Building and Environment, 109453*, 109453.

Yaralıoğlu, K. (2001). Performans değerlendirmede analitik hiyerarşi proses. *Dokuz Eylül Üniversitesi İktisadi İdari Bilimler Fakültesi Dergisi, 16*(1), 129.

Adaptability of UN SDGs by Worldwide Fashion Brands

R. Surjit, M. Shubha Shree, G. M. Indu, and B. Anusri Mahalakshmi

1 Introduction

The fashion industry is one of the largest polluting industries in the world and is also known for its negative impact on the environment, which leads to the question of the fashion industry's sustainability in the future. To promote sustainability and prosperity in the world by 2030, the United Nations in 2015 introduced Sustainable Development Goals. These goals address various aspects of the development of the world. Due to the current fast fashion trend, there is rapid production and consumption followed in the apparel sector, leading to unethical practices being followed at times. To avoid these practices, fashion brands should adopt these SDGs to follow ethical production practices and bring about a positive consumption pattern. These SDGs are required for the betterment of the environment and to decrease the pressure exerted on the environment due to continuous and rapid production carried out to meet the increasing demands. Many fast fashion brands have adopted an endless number of sustainable practices and alternatives to compensate for the depletion actions that they have done to Mother Earth. These practices should be aligned with the SDGs so that some visible improvements toward a sustainable future can be noticed. In today's world, many different brands understand the need to achieve the SDG goals; thus, they are actively indulged in implementing new programs and projects that contribute to and are aligned with the SDGs set by the United Nations. The knowledge of the goals and principles is the primary requirement of a brand or a company, and second comes aligning their goals with the United Nations SDGs, finally working and focusing on the attainment of the goals without any deviations.

R. Surjit (✉) · M. Shubha Shree · G. M. Indu · B. Anusri Mahalakshmi
Department of Fashion Technology, PSG College of Technology, Coimbatore, Tamil Nadu, India
e-mail: rst.fashion@psgtech.ac.in

S. S. Muthu (ed.), *Sustainable Textile and Apparel Chain Management*, Sustainable Textiles: Production, Processing, Manufacturing & Chemistry,
https://doi.org/10.1007/978-3-031-80240-9_4

2 UN SDG Goals

In recent years, deforestation, pollution, loss of biodiversity, and climate change have created a huge impact on the ecosystem, as well as on human health. In the past few decades with the increase in the population of the world, individual consumption has also increased which leads to unsustainable utilization of resources. In September 2015, all the world leaders came together and formed 17 Sustainable Development Goals, which aimed at addressing the main problems in terms of humanity and the planet. These Sustainable Development Goals apply to all countries irrespective of whether the country is developed or not. The main aim is to promote global partnership toward sustainability. The 17 Sustainable Development Goals are listed below:

- No poverty: This goal mainly focuses on eradicating poverty in all its forms. It also focuses on reducing extreme poverty among the people and removing poverty in all its forms for at least half of the population. It addresses the economic, social, and environmental dimensions of reducing poverty. This can be achieved by coordinating with the government, civil sectors, private organizations, and international organizations.
- Zero hunger: This focuses on putting an end to hunger, reducing malnutrition among the people, increasing food security by ensuring that an equal amount of supplies is reaching all the population segments, and promoting sustainable agricultural activities. It also focuses on increasing the productivity and revenue of small-scale producers.
- Good health and well-being: It aims to ensure good health for people at every stage of their life. It involves reducing maternal mortality, putting an end to epidemic diseases, achieving universal health coverage, and reducing deaths caused due to pollution. It addresses both communicable and noncommunicable diseases and ensures equal access to healthcare for all (Howden-Chapman et al., 2017).
- Quality education: It focuses on providing quality education to every person and giving them opportunities for lifelong learning. It states that education plays an important role in developing an equal society (Boeren, 2019). With investment in education, it is easier to create a sustainable society where each individual makes their contribution to the development of the society. It ensures that there is no gender disparity in obtaining education and that everyone has equal opportunities in learning; it also makes sure that it is affordable for everyone.
- Gender equality: It ends all forms of discrimination and violence against women, ensures that women participate in all sectors, and promotes gender equality. It focuses on eradicating child marriages and forced marriages from society. It ensures equal participation of women in all sectors. It is not only considered a fundamental human right but also a foundation for a peaceful and sustainable world.
- Clean water and sanitation: It ensures that every person in the country has access to clean water at an affordable rate and proper sanitation is present. It aims to

improve the quality of water by decreasing the pollution created in the environment, to prevent contamination by building proper sanitation facility, to promote usage of water in a sustainable way to make it available for everyone, and to implement water management systems. With the availability of clean water and proper sanitation, the quality of life is being increased.

- Affordable and clean energy: Its main aim is to give access to affordable, reliable, and modern energy resources to all. It also targets to reduce air pollution by improving the purity of the air for survival and also increasing energy efficiency. Implementing modern technologies and building more infrastructure can pave the path toward more sustainable energy. Thus, with this goal achieved, people will have improved living standards and increased economic growth (McCollum et al., 2017).
- Decent work and economic growth: It promotes sustainable economic growth, ensures that decent work is available for all, and increases the per capita income of the individual, Along with promoting entrepreneurship, increasing the quality of the jobs, and promoting the labor rights for all. It also aims to increase the GDP by at least 7% per year in least developed countries. By achieving this goal, we would create equal and sustained jobs for all (Rai et al., 2019).
- Industry, innovation, and infrastructure: It highlights the necessity of improving the industry, infrastructure, and innovation in technology. Development of a sustainable industry, improving the infrastructure, and increasing innovation play a key role in Sustainable Development Goals (Denoncourt, 2020). It also focuses on increasing the accessibility of finance to the newly built and small-scale industries. It also aims to create jobs for the people and ensure that they get access to basic services.
- Reduced inequalities: It aims to reduce inequalities between other countries and also within the same country. Focus on reducing inequalities in all forms, such as due to age, income, sex, religion, ethnicity, and other factors, and increase the income of the population to a greater extent. By achieving this, we can build a society where there is an improved quality of life for all. It ensures equal opportunities for all and reduces imbalance (Kuhn, 2020).
- Sustainable cities and communities: It mainly addresses the problem of creating housing, transportation, moving to various cities, social participation, and so on. The main aim is to create social cities and communities by improving their quality of lifestyle as well as developing the cities in such a way that any problems that arise in the future can be controlled (Vaidya & Chatterji, 2020).
- Responsible consumption and production: It ensures sustainable consumption of the resources, effective production, and transportation system. It focuses on encouraging the industries to implement sustainable practices. The main aim is to achieve sustainable management and effective usage of natural resources within 2030 and protect the world's culture. With this practice, we can reduce the negative impact on the environment and improve the quality of human life in a more sustainable way (Gasper et al., 2019).
- Climate action: It Focuses on changing climate as it acts as a barrier toward sustainable development worldwide. It also emphasizes the already existing

damage to the environment due to climate change and greenhouse gas emissions (Lofts et al., 2017). Integrating climate measures into policy and promoting education and awareness among people regarding climatic change and their impacts can bring about a good outcome to the environment.
- Life below water: It aims to protect the ocean and marine life by addressing issues such as marine pollution and overfishing as the oceans are required to maintain a balance in the ecosystem. It is also important for economic and social development. Promoting aquaculture as one of the sustainable methods without polluting the ocean, as well as no over-cultivation can help in this regard (Kerton, 2023).
- Life on land: The main goal is to promote sustainable use of the ecosystem, repair the existing damages to the environment, and reverse land degradation (Kerton, 2023). It focuses on promoting sustainable management in all its forms. It focuses on ending poaching, promoting the sharing of natural resources among the countries, and enhancing global contribution.
- Peace, justice, and strong institutions: The main aim is to promote nonviolence and peace among the population, put an end to trafficking in all forms, ensure justice is equal for all, reduce corruption, put an end to all forms of corruption activities, and promote strong institutions. It is one of the most innovative aspects of the Sustainable Development Goals. Peace can be brought into society with all contributions and support from the people's side (Hope, 2020).
- Partnership for the goals: It aims to focus on increasing global partnership for the sustainable development of countries. To strengthen the mobilization of domestic resources, focus on adapting to modern technology, increase the exports of developing countries, and remove the trade barrier for least developed countries. The increase in collaboration with other countries creates the support to achieve sustainable goals (Aroonsrimorakot & Vajaradul, 2016).

These are the 17 important Sustainable Development Goals. Each of these goals has many targets to achieve and is also measured. These goals create a pathway for developing a world that practices sustainability in all its forms.

3 UN SDG Principles

The most basic principle is that no person should be left behind in society and every person should have equal rights in society (Mattera & Alba Ruiz-Morales, 2021). The foundation of these principles is mainly classified into People, Planet, Prosperity, Peace, and Partnership. The agenda for Sustainable Development Document 2030 addresses these 5Ps:

- People: The main aim is to end hunger and poverty in all its forms so that every individual in the world lives a healthy and sustainable lifestyle. The principles aim to make people live their lives to their full potential and equality. Thus, the global communities can come forward and work together to build a world where each individual expresses their full potential.

- Planet: The main determination is to reduce pollution completely and decrease the degradation of the environment by promoting sustainable practices in every form possible so that the current demand is also met without causing damage to the resources available for the future.
- Prosperity: The main determination is to make sure that all the people live a prosperous life in society with harmony. The aim is to increase the economic growth of the world by not causing any damage to the environment . Thus, the environment should be in such a way that prosperity should be shared and it becomes sustainable for the future generation.
- Peace: The determination is to create a peaceful society without any violence and fear and with access to justice for everyone. Any society cannot be made sustainable without peace, and a peaceful society is the right one for sustainable development. Strengthening the laws promoting the rights of humans, and building institutions can bring about peace in the society.
- Partnership: The determination is to create a global partnership between the countries to achieve the Sustainable Development Goals. It calls for mobilization of resources and equal participation of all countries and people. We can create a sustainable future for all by adhering to this principle.

4 Need for UN SDG

The doubt that prevails in everyone's mind: "Can fashion ever be sustainable?" Yes, a difference can be made in fashion and its related sectors through the varied and responsible choices of every individual. The goals help to focus on a particular aspect, to set priorities, and to work toward them without any deviation to achieve. SDGs bring in trust and reputation, which results in mutual benefits. For many business companies or brands, the tag given to them as "SDG fit" adds additional value and draws more sustainable-minded customers as sustainability is gaining more attention in today's resource-decreasing world (Pedersen, 2018). Transparency is a primary aspect that every brand should hold on to as this is considered to be the initial step taken towards change. Brands are expected to be transparent in attributes such as the sourcing methods followed, their stakeholders, tiers of their supply chain, best and worst practices followed, etc. The urgent issues faced by the upcoming brands and industries include male dominance, which adds to gender inequality, living wages provided to the operators, and waste generation, which are left unnoticed as the brands are concerned with improving their production, proper understanding of the SDGs is a need. The SDGs in one or another way help the brands analyze, identify, and thus act according to the issues frequently faced. In today's resource-depleting world, there is an alarming requirement for goals to be set that could help obtain continued contributions from all sectors of the fashion industry.

SDGs bring in a sense of belonging and inclusiveness among the companies, which in turn becomes the driving force to take responsibility. Another aspect that determines the need for SDGs is leadership qualities; companies whose mindset or goals align with the SDGs take leadership and focus on attaining it and also will end up pulling like-minded people into it, which results in a step forward toward achieving the preset goals. Goals set become a source of awareness to common individuals when these are being intensely spoken about by famous brands and companies. With the increasing world population, there is said to be a major increase in the demand for apparel by 63% in developing countries. This in turn will raise the requirement for input materials such as raw materials, water, energy, manpower, etc. (Hansson Reuter & Zetterlund, 2018). To tackle this situation on the sustainable front, aligning and following the SDG goals of the United Nations are crucial.

5 Role of UN SDG in Sustainable Development

The UN Sustainable Development Goals play a vital role in encouraging sustainability and its development across the world. Today's world faces various environmental, economic, and social challenges. These challenges are tackled by UN SDGs, which provide a holistic framework. Many countries have incorporated the UN SDGs into their national policies to match their worldwide sustainable principle toward achieving development in sustainability. The UN SDGs ensure that it covers all the sectors of society, including communities that are vulnerable and underserved. It also ensures that no one is left out from the benefits of the efforts taken. The UN Sustainable Development Goals offer the nations and stakeholders a pathway to work united toward encouraging peace and sustainability, satisfying the current and future needs of the society, and also protecting the earth's natural sources and environment (Saxena et al., 2021).

The role of UN Sustainable Development Goals also focuses on the reduction in inequalities and social cohesion across the world. The goals also promote partnerships with organizations and governments to provide possible solutions to the global issues that affect the sustainable ecosystem on the planet. SDGs support the transition to modern sustainable energy sources that are reliable and affordable. The goals minimize the environmental impact of a product throughout its lifecycle by promoting sustainable manufacturing patterns and responsible consumption. The UN Sustainable Development Goals address international and transnational sustainability problems to face the global concerns affecting the environment. The goals propose certain adaptation strategies for sustainable development in manufacturing industries, which results in a reduction of waste and environmental protection from chemicals used in the industry (Leal Filho et al., 2018).

The role of UN SDGs in the sustainable development of a fashion brand is crucial as the fashion industry is one of the most polluting industries in the world. The value and reputation of a brand is increased by following sustainable measures from the UN SDGs. By implementing UN SDGs in fashion industries, the risks of unsustainable practices and costs associated with it are reduced. The sustainable production methods followed by implementing SDGs gain customers' trust and thereby help in capturing the fast fashion market from competitors. Promoting women's empowerment by employing women in the supply chain and production sections of fashion industries is one of the UN Sustainable Development Goals that would achieve gender equality. The UN sustainable development goals help a fashion brand in gaining a loyal customer base and improving the brand's image (Thakker & Sun, 2023).

6 Benefits and Scope of UN SDGs

The UN SDGs provide the pathway for a better future through the organized efforts of society. It provides a roadmap for addressing the global challenges. The SDGs framework can be a better way to avoid wrong choices between the three domains of sustainable development, that is, economic, environmental, and social. In following the UN goals of sustainability, workers and their families are raised above the poverty line. Corporate social responsibilities are fulfilled and hazardous substances in the production are removed, which in turn promotes a healthier environment. Public awareness of labor rights, health, and safety have been increased in the past couple of years due to the implementation of UN sustainable development goals. Gender equality is promoted and overall women's empowerment is encouraged. Life on land and life below water both have benefited from the goals established. Environmental risk factors such as ozone depletion, greenhouse gas effect, and smog formation due to hazardous chemicals used in production are reduced by following UN sustainable development goals (Thakker & Sun, 2023).

Products that are produced sustainably reduce hazardous effects on the environment and create a balance between natural and manmade sources. The environmental benefit is acquired by the usage of high-quality raw materials, which can be used in the long run. The concept of sustainability increases consumer convenience and societal development. Consumer convenience is achieved using natural resources as materials so that the consumer feels comfortable to use the product over and over leading to sustainable practices. In fashion industries, sustainable production can lead to reduced pollution in the overall physical environment as clothing production contributes more pollution than any other industries. Sustainable material in production reduces overall cost as it can be reused to produce a new product. Research and development cost to develop a new product is eliminated if recycling and upcycling are promoted. Sustainable fashion supports the entire system to reduce the negative effect on society through eco-fashion, ethical fashion, and green fashion by following UN sustainable development goals. All 17 UN SDGs benefit nature, people, and society by all means. As discussed earlier, fashion brands following the UN SDGs benefited in several ways, such as environmental benefits, comfort to the wearer, cost-benefit, and social benefit to the local community (Binti Shafie et al., 2021). UN SDG goals and principles provide a comprehensive approach to challenges that are faced globally. It focuses on improving the well-being of every individual in the world and providing them with a life where all are treated equally.

The UN SDGs are the integrated framework of targets. The future scopes of UN SDGs are vast and complex. The goals ensure that by 2030, all people will experience peace and prosperity in life and the environment. To achieve the goals by 2030, continuous monitoring of progress, adopting techniques, and identifying the challenges are required. Partnerships and collaborations with other industries or governments are crucial in achieving these goals. The promise of sustainable development lies in leveraging advancements in technology and innovation. By embracing renewable energy, digital solutions, and data analytics, the journey toward achieving the SDGs is easier. Population growth may lead to running out of resources shortly. By 2050, the population may rise to 9.8 billion. So, by following the suggested consumption habits and alternative sustainable energy sources, resource depletion can be minimized.

Promising changes in the industries have been noted, including the trend toward sustainability, which has tripled in the last few years. Given that social progress outpaces environmental progress, it's imperative to prioritize initiatives aimed at reshaping power dynamics, reducing inequality, fostering development, and enhancing education in lower-income nations. This focus is essential for safeguarding ecosystem services and advancing overall well-being. The goals that include socio-ecological policies develop sustainability and well-being. In summary, the prospects of SDGs are interlinked and extensive, necessitating endeavors across various domains to achieve a sustainable global community (Henderson & Loreau, 2023).

7 Brands Following UN SDG Goals

Many companies and brands have stepped in to actively participate in the sustainable development of our environment. Top brands are more concerned about aligning their goals with the UN SDGs There are SDG goals that are found to be more connected to the fashion industry globally. Out of 17 goals set, 4 goals are relevant, and these goals are SDG 5: Gender equality, SDG 8: Decent work, SDG 12: Responsible consumption and production, and SDG 13: Climate action. As highlighted in the previous sections, one of the key aspects required in a brand is transparency to its stakeholders. In the Fashion Transparency Index report provided by Fashion Revolution (2019), about 200 brands were analyzed based on a questionnaire sent to them, and while many brands participated, only five had in the range of 61–70% transparency, namely, Adidas, Reebok, Patagonia, Esprit, and H&M; they disclosed some aspects in which they assess their supplier, list of their go-to suppliers, an initiative taken for the betterment of their workforce, human rights and environmental policies followed, etc., and unfortunately, none of the brands came under the transparency range of 70–100% as some details about attributes such as the gender pay gap, addressing of women's issues, rate of carbon emission, and water utilization were not disclosed in detail.

7.1 Luxury Fashion Brands and SDGs

Luxury brands are focused on quality over quantity, which makes their product last longer than the ones that are produced in bulk quantities; moreover, these luxury products are further reused, repurposed, or even recycled again and again, limiting the utilization of new resources. This contributes to SDG 12: Responsible production and consumption. These brands make sure that they provide their laborers with safe conditions to work and that they are paid accordingly. This takes the brand closer to SDGs 1, 3, and 5. Luxury brands that are focusing to align with SDGs are:

7.1.1 Chanel

As an independent house of luxury, Chanel has been a brand that focuses on not only the longevity of their house but also the planet's longevity. They have developed an eco-designed product line that aligns with the UN SDGs. This product line has taken into consideration that all the raw materials and the processes followed in the lifecycle of the product, from sourcing to its packaging and its afterlife, are focused on reducing carbon impact. They have declared a ban on the use of fur and exotic hides, thereby protecting the wild from being hunted. They have also tied up with suppliers who are indulged in sustainable approaches, such as "Sulapac," a startup that produces biodegradable plastic alternatives, and "Evolved by Nature," which produces nontoxic silk alternatives. They have set their goals in the year 2024 to achieve net zero greenhouse gas emissions by the year 2040, which aligns well with the SDGs.

7.1.2 Stella McCartney

It is a brand that is forward in terms of sustainability by innovating new alternatives such as Savian (a plant-based, plastic-free alternative for fur), which is found to produce 40–90% less CO_2 than conventional furs. It is called NATIVA wool (regenerative wool). They have a partnership with Veuve Clicquot, by which the waste of grapes is turned into a vegan luxury, a better alternative for animal leather. This leather is free from toxic solvents and metals. Another raw material supplier, Biopuff, provides a regenerative alternative for feather down; these are aligned with SDG 12, 13, 14, and 15. This brand has been a member of the Ethical Trading Initiative, which focuses on the betterment of its workers and aligns with SDGs 3, 4, and 8.

7.1.3 Burberry

This brand is concerned about the services, products provided, and packaging. The services provided by them that align with the SDGs are aftercare and circular services for the products that have reached the end of their lifecycle, which includes upcycling, refreshing, and repairing. This prolongs the lifetime of the product and also prevents it from ending in landfills before being utilized to its fullest. They are committed to using responsibly sourced raw materials: The main material used in all garments contains 70% organic content or 50% recycled content. A dedicated innovation team is available to develop better alternatives and next-gen materials. To reduce the carbon impact, they have incorporated packaging that is 100% free from plastic.

7.2 *Fashion Brands and SDGs*

Other than luxury fashion brands, many other brands are also aligning their products and processes with SDGs, and the details of them are given below:

7.2.1 Gant

A clothing and accessories Swedish brand having roots in America, Gant believes consciously and responsibly in producing products. They follow seven rules for product care or wardrobe maintenance that include Repair, Refresh, Relove, Remake, Regive, Recycle, and Rent (Gant, 2024). These rules primarily revolve around the concept of circularity and are given in Table 1.

Table 1 Seven rules of sustainability followed by the brand Gant

Repair	Little mending can be done to overcome wear and tear and extend the lifespan of the product.
Refresh	The right care for the right garment and proper storage make the garment last longer.
Relove	Their garments are more durable, so they ask their customers to reuse the garments as "Reloved."
Remake	Limiting the waste produced and utilizing every bit of scrap to produce limited edition collections.
Regive	As a commitment to the UN SDG goals, they provide clean water and sanitation to some communities in parts of Bangalore, India.
Recycle	As a key to sustainability, they encourage their customers to recycle their used garments instead of letting them into landfills.
Rent	Partnership with an online rental service named HURR to enable their customers to get garments for occasional wear.

7.2.2 Nike

This sportswear brand is the only fashion brand among the seven brands that are fast approaching the sustainable deadlines of the United Nations that are to be met by 2030. They are aiming to protect the environment through zero waste and zero carbon emissions. This is aligned directly with the twelfth SDG, which focuses on responsible consumption and production.

7.2.3 Zara

Zara is the leader of fast fashion as they have been the brand that produces products and makes them available in the market for their customers in a very short period. Zara, though being the top fast fashion brand, also paved the way towards sustainability, and they have taken up responsible practices (Dockrill, 2022). One among these is "closing the loop," a program for circular fashion, where their customers can give all the garments that they feel will not be used again. These garments are sent for recycling rather than ending up in landfills. Additionally, what makes them unique is that they take up products from any brand. Apart from this, they also use bags that are recycled. Zara is known for their transparent activities. They have written and published many reports regarding their supplier relations (Jha & Veeramani, 2021).

7.2.4 TOMS

TOMS, a footwear company, is well known for its campaign called "One for one," where when a customer purchases a pair of shoes, another pair is sent to a needy child. They are also focused on equity, employee empowerment, and anti-racist initiatives. About one-third of their profits are spent for good, and this is one of the brands that focuses on the mental health and wellbeing of individuals.

7.2.5 Patagonia

An American company known for its famous outdoor clothing, this brand abides by the SDGs by following some unique aspects, such as the nonuse of polyfluorinated chemicals (PFCs). This brand produces water-repellent outdoor clothing. In the recent collection of spring 2024, 94% of their water-repellent chemical solution is made without hazardous chemical components such as PFAs and PCAs. It also follows a concept called the fair trade movement as it is the founding member of the Fair Labor Association. With this practice in place, Patagonia tries to ensure that their apparel assembly factories are paying their workers a living wage on average.

They have succeeded in this initiative in 10 out of 29 factories, amounting to 34%, and this aligns with SDG 8.

7.2.6 Levis

This brand, abiding by the SDGs, has implemented a lot of practices and is actively contributing to achieving the SDGs set. In par with SDG 3, they are more concerned about their workers and employees. According to the SDG report released in 2021, they focus on their employees' well-being by providing financial assistance when required, global insurance, work-life program (Employee Assistance Program), counselling, paid family leave, and ensuring mental well-being of their employees. They have also developed a process for water use, through which the water conservation increased. Between 2011 and 2021 they saved around 11.5 billion liters of water by recycling, thus effectively abiding by the SDG 6—Ensuring availability and sustainability of water and sanitation to all. Out of 17 sustainable development goals set, Levis has focused on and incorporated many projects that aim to achieve 10 sustainable goals.

7.2.7 Uniqlo

Uniqlo belongs to a fast retail fashion company in Japan. They have contributed to sustainability by practising circular fashion, in which their products are recycled and prevented from being thrown as waste. In the 2022 spring and summer, Uniqlo started incorporating recycled plastic collected from bottle waste. They have partnered with similar mindset entities such as Toray, who produce the Dry EX polo shirts. Uniqlo collects the waste and used garments from their customers and sends them for recycling. Out of the garments that are collected from the customers, those found in good condition are donated to the underprivileged people of the economy, and the rest are sent for recycling (Zahari, 2021).

7.2.8 H&M

H&M (2024) believes in satisfying the needs in such a way that it is economically, socially, and environmentally sustainable; the goals that they have set are meant to achieve this. They are a company that aims to achieve net zero climate impact by incorporating circular models for the product lifecycle and supply chain. They value their workers and respect them, and they are the supporters and promoters of diversity in every aspect. The climate goals set include reducing $CO2$ emissions by 56% by the year 2030. The progress in the climate goal till 2023 is estimated to be a 24% reduction in scope 1 and 2 emissions and a 22% reduction in scope 3 emissions. They practice ethical sourcing, for which they have set goals in which 100% of the materials sourced for the commercial purpose should be either recycled or sustainably sourced by 2030, progress shown by 2023 is that they have achieved it by 85%. The next goal set to be achieved by 2030 is making all the packaging from recycled or sustainably sourced materials which has shown progress to 79% in 2023. To abide by the vision to be fair and equal, the goal that they have set is to increase the percentage of female supervisors and female representatives in the tier 1 supply chain of production, which has shown a progress of 27% and 63% by 2023, respectively.

7.2.9 Decathlon

Decathlon is engaged to reduce the absolute scope 1, 2, and 3 greenhouse gases by 42% by 2030 to decarbonize the activities of the company by reducing the carbon intensity in the product, preferring and utilization of sustainable raw materials, and encouraging suppliers to utilize the renewable resources over the nonrenewable ones. They are focused on sustainable production by optimizing the design of the products, which helps in maximizing the economic value of the product (products that are durable, recyclable, and repairable). They focus on scaling up their business as circular models, thereby reducing the impacts and favoring the environment (Decathlon, 2024).

7.2.10 Everlane

Everlane is an online fashion retailer brand that continuously interacts with its customers to be a key participant in their business activities and to raise awareness among them about environmental and social issues. They majorly focus on three key sustainable practices, which include:

- Keeping the planet clean by utilizing very limited inputs, such as reducing the natural resources consumption and producing output that are cleaner, which is achieved by minimizing the waste emitted, and pollution by emphasizing safer chemistry in the overall value chain.
- Committed to a reduction in emission of greenhouse gases by more than 50% per product by 2030 and zero emissions by or before 2050.
- Do right by people, which focuses on achieving gender equality and fair living wages, thereby enhancing the livelihood of the workers (Everlane, 2024).

From the above practices and goals set, Everlane is said to be aligned with Sustainable Development Goals 1, 5, 6, 7, and more.

7.3 *Common Measures Taken to Achieve SDG*

It is said that around 1% of the textile goods in the world are being recycled, 25% of the newly manufactured garments are not sold, and 35% of the microplastics from synthetic textiles are leached into the water bodies (Thakker & Sun, 2023). On the whole, many brands and industries belonging to the fashion and textile sectors are actively engaged in meeting the sustainable development goals set by the United Nations. All the brands have contributed and are continuing to contribute to the set goals. To meet every goal, some common measures can be implemented by all the global brands. By implementing circular economy practices, we can implement SDG in the fashion sector, encouraging customers to return used clothes that can be used for upcycling and repurposed by industries (Cai, 2021). Using renewable energy sources and reducing the carbon footprint leads towards efficient usage of energy and thus results in reduced emissions. Responsible production and consumption practices are to be followed in the manufacturing industries as well as fashion industries (Gardetti & Muthu, 2020). Creating consumer awareness about SDG through magazines and online platforms can be done. Implementing circular-based manufacturing can help to reduce waste generation, and collaboration with stakeholders can lead to industry-wide changes. Industry can also conduct sustainability assessments to know about the impact created by them on the environment and can improve their contribution to the well-being of society. In these ways, fashion industries can implement Sustainable Development Goals to contribute to a sustainable future.

Thakker and Sun (2023) did an extensive study and stated the issues and measures to be followed for achieving the SDG's in apparel and textile industries. The common measures stated by them to achieve SDG are detailed in Table 2.

8 Conclusion

The fashion industries should start following and adapting to the Sustainable Development Goals to ensure responsible production and consumption practices. Providing awareness to the consumers and adopting the best practices in the industries can reduce the negative impact on the environment to a great extent. The commitment given by fashion brands towards sustainability will be a crucial driving force in accomplishing the United Nations's vision for a sustainable future. SDGs are not new, but they are the old sustainable agenda that is made easier to understand and implement. Though fashion is back in terms of sustainability, raising awareness has let fashion and its related industries take a step into sustainability. Today's luxury brands are more concerned about the environment and are taking measures to meet the SDGs. This makes them stand out and become role models for the developing, upcoming brands. Businesses are forced and pressured to incorporate environmental responsibilities, but these tend to be the driving force that increases competition

Table 2 Measures taken to achieve SDG

SDG	Measure
1. No poverty	To provide financial support for the farmers who produce cotton, flax, and bamboo and lift them above the poverty line
2. Zero hunger	To provide food free of cost for all the workers and operators in the company To educate the workers about the importance of nutrition through different kinds of food and the amount of food to consume to overcome malnutrition
3. Good health and well-being	Implementing projects and programs that reduce or eliminate the use of harmful chemicals For the betterment of the workers, encouraging them to indulge in physical activities by having a specific time allocated for exercise during working hours
4. Quality education	To impart knowledge to the workers and operators on the technical aspects and overview of the supply chain Also inculcating the basic knowledge of sustainability and its importance in today's world to the workers Providing funds and an additional percentage of money to the economically backward operators for educating their children
5. Gender equality	Providing equal wages and opportunities for both male and female workers. Prohibiting child labor Employment of women in leadership and managing roles
6. Clean water and sanitation	Implementing a wastewater recycling plant and making sure that waste water is not ejected out of the factory without treatment Ensuring sanitized drinking water all over the factory
7. Clean energy	Utilizing energy by implementing solar plants and other sustainable sources of energy The textile processes carried out should not harm the environment
8. Decent work and economic growth	The mental and physical health of the workers should be taken care of Ensuring to contribute to circular fashion rather than dumping them in the garbage
9. Industry, innovation, infrastructure	To build an organized and safe infrastructure with all the safety measures incorporated according to the rules set The new projects and innovations carried out should meet the requirements of the sustainable development goals Improving the infrastructure and ensuring all the safety measures are implemented
10. Reduced inequalities	Aims to control inflation and reduce inequality in promotion and wages
11. Sustainable cities and communities	Taking up corporate social responsibilities by the brands The upliftment of the people who belong to the lower communities of the society
12. Responsible production and consumption	Encouraging the customers towards minimalism and providing awareness to purchase only according to their needs Taking up only make-to-order rather than make-to-stock production Use of biodegradable materials as much as possible for packaging and designing aspects

(continued)

Table 2 (continued)

SDG	Measure
13. Climate action	The main focus is on reducing or even eliminating carbon emissions, by controlling the release of greenhouse gases, thereby reducing the factors responsible for global warming
14. Life below water	Installing an effluent treatment plant for all the factories. Protecting the aquatic life by treating the effluent waste before emission and recycling the released wastewater again
15. Life on land	Holding on to circular fashion and cradle-to-cradle concept to prevent the garments from being dumped in the landfills To plant more trees such that it compensates the destruction done to Mother Earth
16. Peace, justice, and strong institution	Brands and industries to indulge in ethical practices that cause less harm to people, animals, and the environment. The brands to invest more in circularity To not compromise on quality and to follow transparent business practices
17. Partnership for the goals	To partner with all the stakeholders that have similar mindsets in achieving sustainable development goals All the textile manufacturers, designers, and consumers work in close collaboration with each other

between two brands or businesses. SDGs are said to provide a stronger bond between three sectors: the policymakers, the civil society, and the private sectors. Achieving the Sustainable Development Goals in 2030 would be a great success for the betterment of tomorrow's globe, this can be achieved only when every individual brand takes this seriously and contributes equally. Implementation of new methods to abide by the SDGs not only helps the betterment of the globe but also adds value to the brand. This becomes an aspect that differentiates the brand from others and attracts customers who are environment conscious.

References

Aroonsrimorakot, S., & Vajaradul, Y. (2016). UN sustainable development goals: 17 aspects for the future world. *Interdisciplinary Research Review, 11*(3), 1–7.

Binti Shafie, S., Binti Kamis, A., & Bin Ramli, M. F. (2021). Fashion sustainability: Benefits of using sustainable practices in producing sustainable fashion designs. *International Business Education Journal, 14*(1), 103–111.

Boeren, E. (2019). Understanding Sustainable Development Goal (SDG) 4 on "quality education" from micro, meso, and macro perspectives. *International Review of Education, 65*, 277–294.

Cai, Y. (2021). *Achieving sustainable development goals (SDGs): Sustainable supply chain management in the fashion industry.*

Decathlon. (2024). *Ecodesign*, [Online]. Available from https://sustainability.decathlon.com/its-because-we-design-our-products-that-we-can-ecodesign-them. Accessed on 20 July 2024.

Denoncourt, J. (2020). Companies and UN 2030 sustainable development goal 9 industry, innovation, and infrastructure. *Journal of Corporate Law Studies, 20*(1), 199–235.

Dockrill, M. (2022). *How ethical is Zara*, [Online]. Available from https://goodonyou.eco/how-ethical-is-zara. Accessed 20th June 2024.

Everlane. (2024). *Fashioning a better future*, [Online]. Available from https://www.everlane.com/sustainability. Accessed 20th June 2024.

Gant. (2024). *Sustainability at Gant*, [Online]. Available from https://gant.com/sustainability/7rules. Accessed 20th June 2024.

Gardetti, M. A., & Muthu, S. S. (Eds.). (2020). *The UN sustainable development goals for the textile and fashion industry*. Springer.

Gasper, D., Shah, A., & Tankha, S. (2019). The framing of sustainable consumption and production in SDG 12. *Global Policy, 10*, 83–95.

H&M Group. (2024). *Goals and ambition*, [Online]. https://hmgroup.com/sustainability/leading-the-change/goals-and-ambitions/. Accessed 20th June 2024.

Hansson Reuter, M., & Zetterlund, L. (2018). Managing for sustainable fashion-A comparative review of the Higg Index 2.0 Brand and Retail Module and the UN Sustainable Development Goals.

Henderson, K., & Loreau, M. (2023). A model of Sustainable Development Goals: Challenges and opportunities in promoting human well-being and environmental sustainability. *Ecological Modelling, 475*, 110164.

Hope, K. R., Sr. (2020). Peace, justice and inclusive institutions: Overcoming challenges to the implementation of Sustainable Development Goal 16. *Global Change, Peace & Security, 32*(1), 57–77.

Howden-Chapman, P., Siri, J., Chisholm, E., Chapman, R., Doll, C. N., & Capon, A. (2017). SDG 3: Ensure healthy lives and promote wellbeing for all at all ages. In *A guide to SDG interactions: From science to implementation* (pp. 81–126). International Council for Science.

Jha, S. K., & Veeramani, S. (2021). Sorting responsible business practices in fast fashion: A case study of Zara. *Journal of Management and Public Policy, 12*(2), 54–58.

Kerton, F. M. (2023). UN Sustainable Development Goals 14 and 15–Life below water, Life on land. *RSC Sustainability, 1*(3), 401–403.

Kuhn, H. (2020). Reducing inequality within and among countries: Realizing SDG 10—A developmental perspective. *Sustainable Development Goals and Human Rights, 5*, 137–153.

Leal Filho, W., Azeiteiro, U., Alves, F., Pace, P., Mifsud, M., Brandli, L., Caeiro, S. S., & Disterheft, A. (2018). Reinvigorating the sustainable development research agenda: The role of the sustainable development goals (SDG). *International Journal of Sustainable Development & World Ecology, 25*(2), 131–142.

Lofts, K., Shamin, S., Zaman, S. T., & Kibugi, R. (2017). Brief on Sustainable Development Goal 13 on taking action on climate change and its impacts: Contributions of international law, policy and governance. *McGill Journal of Sustainable Development Law, 13*, 183.

Mattera, M., & Alba Ruiz-Morales, C. (2021). UNGC principles and SDGs: Perception and business implementation. *Marketing Intelligence & Planning, 39*(2), 249–264.

McCollum, D., Gomez Echeverri, L., Riahi, K. and Parkinson, S., 2017. Sdg7: Ensure access to affordable, reliable, sustainable and modern energy for all.

Pedersen, C. S. (2018). The UN sustainable development goals (SDGs) are a great gift to business! *Procedia CIRP, 69*, 21–24.

Rai, S. M., Brown, B. D., & Ruwanpura, K. N. (2019). SDG 8: Decent work and economic growth–A gendered analysis. *World Development, 113*, 368–380.

Revolution, F. (2019). *Fashion transparency index* [Online]. Available from https://www.modasosteniblebcn.org/wp-content/uploads/2020/05/Fashion-Transparency-Index-2019.pdf. Accessed 20th June 2024.

Saxena, A., Ramaswamy, M., Beale, J., Marciniuk, D., & Smith, P. (2021). Striving for the United Nations (UN) sustainable development goals (SDGs): What will it take? *Discover Sustainability, 2*, 1–14.

Thakker, A. M., & Sun, D. (2023). Sustainable Development Goals for Textiles and Fashion. *Environmental Science and Pollution Research, 30*(46), 101989–102009.

Vaidya, H., & Chatterji, T. (2020). *SDG 11 sustainable cities and communities: SDG 11 and the new urban agenda: Global sustainability frameworks for local action* (pp. 173–185). Towards sustainability science, policy, education and practice.

Zahari, Z. A. B. (2021). *Case Study of UNIQLO; Sustainability in Fast Fashion and Consumers' Awareness*. E-Portofolio Jurnal Teknologi UTM.

Challenges for the Achievement of the Sustainable Development Goals in the South American Andean Camelid Chain

Rosa Patricia Larios-Francia and Oscar Cárdenas Minaya

1 Introduction

The year 2024 has been declared by the United Nations as the International Year of Camelids (FAO, 2024); Camelids are considered a key livelihood resource for millions of families living in harsh habitats in more than 90 countries, particularly among Indigenous Peoples and local communities. Camelids, ranging from alpacas to Bactrian camels, dromedaries, guanacos, llamas, and vicuñas, contribute to food security, nutrition, and economic growth, as well as having great social and cultural relevance for many communities around the world. Activities around the camelid value chain play an important role in progress towards the Sustainable Development Goals (SDGs) related to eradicating extreme poverty (SDG 1), fighting hunger (SDG 2), ensuring decent work and economic growth (SDG 8), generating spaces for technology transfer to improve the conditions for innovation and infrastructure in the chain (SDG 9) and the sustainable use of terrestrial ecosystems (SDG 15) (United Nations, 2015) are a priority to ensure the livelihoods of the South American camelids value chain.

The breeding of South American camelids constitutes the only form of productive use of the extensive areas of natural pastures where agriculture or the breeding of other domestic species is almost impossible. It also provides breeding communities with products such as meat and fiber and byproducts such as hides, leather, and milk and serves as a means of transporting goods and people, provides biological

R. P. Larios-Francia (✉)
Universidad del Pacífico, Lima, Peru
e-mail: Rp.lariosf@up.edu.pe

O. Cárdenas Minaya
Instituto Nacional de Investigación Agraria, Puno, Peru

S. S. Muthu (ed.), *Sustainable Textile and Apparel Chain Management*, Sustainable Textiles: Production, Processing, Manufacturing & Chemistry,
https://doi.org/10.1007/978-3-031-80240-9_5

fertilizer, and is well adapted to places where other livestock species would not survive (FAO, 2005a, b, 2008, 2018).

2 Background

2.1 South American camelids

The origin of South American camelids dates from five to ten million years ago, in the late Miocene stage; the genera that exist in the South American fauna are Paleolama and Lama, which appeared in the South American Pliocene, differentiated by their size, with the giant Paleolama and the Lama being slightly larger than the current llama and another smaller one called Pro-vicuña.

The genus Lama appears later than Paleolama, almost at the end of the Pleistocene, differentiating two species, Lama guanicoe and Lama vicugna, between 400,000 and 500,000 years ago. As a result of their domestication, Lama glama and Lama pacos emerged, which existed for the last ten thousand years, receiving the action of man through interspecific crossbreeding, which gave rise to the great variation that exists today (Huanca Mamani, 2020).

According to the Andean vision, the origin of the South American camelids was a gift from the god Pachacamac, who gave them to men who suffered hardship because they did not have enough food and clothing. The llamas and alpacas were given to them as a heritage on the condition that they would conserve them, take care of their reproduction and protect them from their natural enemies and diseases. This legend is considered to be one of the main reasons why the Indians offered great resistance when the Spaniards demanded their exchange with the sheep, and that is when they spread out to the remote places of the Puna, where the Spanish could not reach and where they live to this day (Huanca Mamani, 2020; T. Huanca, 1996; W. Huanca, 2013).

South American camelids are represented by four species, two domestic species, the alpaca (Vicugna pacos) and the llama (lama glama), and two wild species, the vicuña (Vicugna vicugna) and the guanaco (lama guanicoe), which live in high Andean areas 3800 meters above sea level.

The breeding and production of domestic South American camelids is one of the most important productive and economic activities on which 70–80% of annual household income in the high Andean zone depends according to FAO (FAO, 2008; Ministerio de Desarrollo Agrario y Riego, 2021a, b).

Peru has the largest camelid population in the world, with more than 4.5 million camelids of the four species (El Peruano, 2024).

2.1.1 The Vicuña

In the case of wild species, the "vicuña", since the Spanish conquest, has suffered dramatic losses in numbers due to indiscriminate hunting, falling from two million to less than 10,000 in 1964. The high price of vicuña fleece on the international luxury goods market led to the killing and over-exploitation of this species, which brought vicuñas to the brink of extinction (Arzamendia & Vilá, 2012; Vilá & Arzamendia, 2022b).

The vicuña is currently categorized in Peru as a "Near Threatened" species and is also listed in Appendix II of the Convention on International Trade in Endangered Species of Wild Fauna and Flora (CITES); more than 50% of the world's population, around 218,000 specimens, are found in Peru (Servicio Nacional Forestal y de Fauna Silvestre, 2023).

After 30 years of protection and effective management, such as the 1979 International Convention for the Conservation and Management of the Vicuña, the vicuña population has recovered, making sustainable management projects viable for the benefit of the Andean people. With a world population of approximately half a million vicuñas, management plans conceived within the paradigm of conservation through sustainable use have been developed in the last decade. See Fig. 1.

As indicated by Acebes et al. (2018), today, live shearing of vicuñas is the only legal way to harvest vicuña fleece and is widespread throughout the Andes, albeit with great variability in techniques and, therefore, different degrees of success in terms of sustainability goals (Vilá & Arzamendia, 2022a, b).

Fig. 1 Vicuña

The vicuña is covered with an extraordinarily soft and shiny coat, its fiber is the finest in the world, ranging from 10 to 15 microns in diameter. The annual production of sheared vicuña fiber is approximately 200 grams per animal, making it an exclusive fiber due to its reduced production worldwide, which is used in the production of luxury clothing.

2.1.2 Guanáco

Furthermore, the "guanaco" is categorized as a "critically endangered" species, due to its small population of approximately 3810 specimens, and has been included in Appendix II of the Convention on the Conservation of Migratory Species of Wild Animals (CMS) (Agencia Agraria de Noticias, 2024).

This species was exterminated by European colonizers at the end of the nineteenth century to give priority to sheep breeding. In the Argentinean Andes, projects of great value to the conservation and sustainable development objectives of capture and shearing have emerged as conservation strategies for the sustainable use of guanacos (Vilá et al., 2020, 2021).

Guanaco fiber possesses qualities that, to the touch, demonstrate its high quality. Another notable characteristic is the reddish hue of its fiber. The guanaco is the only South American camelids that can easily adapt to different environments (Fig. 2). It can live at sea level or at altitudes of up to 4600 meters above sea level (Grupo Michell, 2024).

Historically, wild camelids have been used by Andean human groups since the first settlements in the Americas more than 11,000 years ago, as cited by Yacobaccio

Fig. 2 Guanaco

(2009). Archaeological evidence shows that the contributions of vicuñas and guanacos to the Andean peoples who were the first American hunters were their meat, skin, fur, and even their bones for making tools (Vilá & Arzamendia, 2022a).

2.1.3 The Llama

The llama (Lama glama) is a strongly built species and the largest of the South American camelids (Fig. 3). The llama has been domesticated and used as a beast of burden for transport since pre-Inca times and even today forms part of many of the diverse customs and religious rituals of Andean peoples.

There are two species of llama: the "Qara" or "hairless," which has developed with little fiber on the body, lacking fiber on the face and legs, with an estimated 70% of llamas being of this type, which is the strongest and is used for transporting cargo, and the "Chaku" or "woolly," which has more fiber on the body, with a dense fleece, composed of fine fibers and of greater length (CITE, 2013). The llama's hair is coarser and coarser than that of the alpaca, but if removed, it can be just as soft (Grupo Michell, 2024). The total world llama population reaches approximately 3.321 million head; Bolivia being the main producer with 60.88%, followed by Peru with 32.52%, Argentina with 4.05%, and in smaller quantities Chile and Ecuador (MIDAGRI, 2020, 2022) (MIDAGRI, 2020). In Peru, the llama population is estimated at 1'104,000 head (CITE, 2013).

Fig. 3 Llama

2.1.4 The Alpaca

The alpaca (Vicugna pacos) was domesticated 6000 years ago in the central highlands of Peru, its breeding was taken to the inter-Andean valleys 3800 years ago, according to archaeological evidence. It is one of the South American camelids species of the Andean biodiversity that has a coat classified as a luxury fiber; considered of high textile value due to its softness, length, fineness, and warmth; presenting a range of 22 natural colors and more than 64 intermediate shades (Huanca Mamani, 2020; Huanca et al., 2011).

The alpaca is considered a sustainable animal because they don't damage the land. Alpacas nibble only the tops of grasses and plants; they do not rip plants out of the ground, resulting in less disturbance of the vegetation and allowing it to grow black. Also, the natural habitat of alpacas is about 3.800 meters above sea level. At this altitude, the water supply is natural, and the land is generally not suitable for agriculture. This makes alpacas more environmentally friendly than all other fiber-producing livestock that often contribute significantly to serious environmental problems (Grupo Inca, 2023).

The world population of alpacas exceeds six million, with 87% concentrated in the Andean region of Peru (El Peruano, 2023). Peru has approximately 4.3 million alpacas between the Suri and Huacaya breeds (MIDAGRI, 2022). See Fig. 4.

In Peru, there are around 92,700 agricultural units of alpaca producers dedicated to the production and breeding of alpacas (MIDAGRI, 2023), which constitute an important source of employment and income from the sale of fiber, meat, and breeding animals. However, more than 76% of camelid producers and breeders are in conditions of poverty and extreme poverty and are considered a highly vulnerable population, even more so due to the effects of climate change (Cárdenas Minaya, 2020).

Fig. 4 Alpaca

3 Effects of Climate Change

Andean pastoralism, like other pastoral systems around the world, is under stress due to climate change, water stress due to glacier melt, major droughts, prolonged snowfall, high solar radiation affecting the Andean ecosystem, land tenure regimes, pressures to become sedentary, difficulties interacting with market economies, isolation, and youth outmigration, as shown in Fig. 5 (Dupuits, 2021; Lanegra, 2021; Postigo, 2009; Vilá & Arzamendia, 2022a).

Climate change is currently a factor that has a negative impact on the productivity of the alpaca sector, degrading the land, and reducing the quality and nutritional potential of natural pastures, due to water stress, which causes an imbalance in nutrients that leads to a decrease in the productive and reproductive rates of the animals, and has also generated annual cyclical losses of thousands of camelids, loss of biodiversity, threatening the sustainability of alpaca activity and the economic livelihood of the communities (Australian Alpaca Association, 2023; Cárdenas Minaya, 2020; FAO, 2018; Ministerio del Ambiente, 2023).

The production system of domestic South American camelids is extensive grazing, based on the forage supply of natural pastures or rangelands; the degradation of the Andean ecosystem is a complex problem that is related to the organizational model of the communities and the management of the land. Likewise, in animal production and management, a predominantly traditional technology is applied with little adoption of technological innovations; there is the existence of intermediation aimed at supplying the external market, which generates price instability and leads to inequitable distribution along the chain, with the producers being the least favored.

Fig. 5 Alpaca herd in frosty weather. (Fuente: (Cárdenas Minaya, 2020))

Communities that breed camelids occupy a lower position in the Human Development Index (HDI), which considers life expectancy and education, generating disparity with populations of similar income (SDG 1), climate change also affects the health of producers, with emphasis on children and older adults, as well as their income is affected by animal mortality or diseases that generate less production of fiber or meat and even the mortality of their species (Aguilar Calla & Torres Zúñiga, 2015; Aguilar et al., 2014; Ministerio del Ambiente, 2023).

The implementation of actions with clear targets for the sustainable development of this population of communities and settlers, related to the end of poverty (SDG 1), ensuring decent work and economic growth (SDG 8), generating spaces for technology transfer to improve the conditions of innovation and infrastructure of the chain (SDG 9), and achieving results in the reduction of inequalities (SDG 10) are a priority to ensure the subsistence of the South American camelid value chain.

4 Strategies for the Achievement of the Sustainable Development Goals in the South American Andean Camelid Chain

In response to the serious repercussions of climate change on the health of camelid herds, which generate a high mortality rate, especially in offspring, by increasing infectious and nutritional diseases and by failures in the transfer of maternal immunity, which limits the productivity of high Andean communities, reducing their income, the National Institute of Agrarian Research (INIA) developed the project "Evaluation of the Effect of Supplementation with Immunomodulators on the Morbidity and Mortality of Young Alpacas," which carried out a follow-up phase of live weight at birth, showing that the newborns of first-time alpacas have a lower birth weight (5.11 kg) than that of multiparous mothers (5.77 kg). This weight would be influenced by the effect of climate change on the nutritional potential of natural grasslands in the dry Puna, as evidenced in Fig. 6 (Cárdenas Minaya, 2020).

Another measure developed and analysed by the project was the effect of colostrum ingestion in neonates, which ensures the amount of IgG immunoglubilins that generate immunological protection to the offspring, these studies and advances in the offspring care process is an action corresponding to SDG 9.

The development of adequate sheds to reduce the effects of frost, hailstorms, and torrential rains is a vital strategy. These sheds should be structures built with materials from the same area, such as stone, mud, wood, and ichu; they should have sloping roofs and adequate windows to promote ventilation; these have proven to be effective in reducing mortality mainly of the young, as they generate an adequate microclimate that favors the shelter of the young, preventing chilling and the presentation of pneumonia (SDG 1, SDG 8 and SDG 9) (Aguilar Calla & Torres Zúñiga, 2015; Aguilar et al., 2014; Huanca Mamani, 2020; Huanca et al., 2018).

Another strategy has been the development of thermal waistcoats and waterproof covers with synthetic material inside to counteract the drop in temperature and the generation of respiratory diseases (Cárdenas Minaya, 2020). See Fig. 7.

Fig. 6 Experimental treatment INIA. (Fuente: (Cárdenas Minaya, 2020))

Fig. 7 Camelids waistcoats. (Fuente: (Cárdenas Minaya, 2020))

The use of natural immunomodulators has also been implemented, with probiotics being considered to have a beneficial effect on disease prevention and animal morbidity and mortality. Plants such as Echinacea purpurea have been used for this purpose, as well as vitamin E intake.

Projects such as Ayninacuy of the Ministry of the Environment have implemented actions with an impact on strengthening livelihoods, including activities for the protection and health of animals, as well as water management and human health care. To improve alpaca nutrition, improved pastures were planted, and the use of high-altitude forage cereals was introduced, and regarding water management, rustic water channels were improved and reservoirs with sprinkler irrigation systems were installed, all in line with SDG 2 and SDG 15 (Ministry of the Environment, 2023).

At a global level, different efforts are being developed, such as the case of the Australian Alpaca Association, which in its concern to generate a sustainable chain is developing the Sustainability Framework, a policy document with recommendations on good practices in terms of animal welfare, biosecurity, environmental management, and conditions of workers in alpaca communities and with proposals for improving the profitability of producers (Australian Alpaca Association, 2023), which will generate impacts in relation to SDG 1, SDG 2 and SDG 15.

All these initiatives and projects should be socialized with all the actors of the South American camelids value chain in the different Andean regions, this will bring an improvement in the current conditions of the Andean communities and in the improvement of welfare and good production and breeding practices.

Each of the activities contributes to the achievement of the different Sustainable Development Goals by improving the conditions of the population, biodiversity, and the ecosystem.

5 Conclusions

The study concludes that the sustainability of the South American camelids value chain is vital for the achievement of the Sustainable Development Goals in high Andean communities. Despite significant challenges, such as the negative impact of climate change due to frost, water stress, loss of quality of grazing pastures, and the economic vulnerability of producers, current initiatives show that it is possible to improve living conditions and protect these ecosystems. It is necessary to continue researching, with the objective of developing new technological initiatives, as well as establishing policies that ensure investment in research, technological development to generate innovation, as well as technology transfer to the people and communities that produce the South American camelids value chain. It is crucial to continue developing and implementing innovative and sustainable strategies that include technology transfer and the adoption of best practices. These actions will not only promote and ensure the economic and social development of communities but also contribute to the conservation of biodiversity and environmental balance in the region. Collaboration between governments, research institutions, academia and local communities is essential to ensure a sustainable future for the South American camelids chain.

References

Acebes, P., Wheeler, J., Baldo, J., Tuppia, P., Lichtenstein, G., Hoces, D., & Franklin, W.L. 2018. Vicugna vicugna. The IUCN Red List of Threatened Species 2018: e.T22956A18540534. http://dx.doi.org/10.2305/IUCN.UK.2018-2.RLTS.T22956A18540534.en

Agencia Agraria de Noticias. (2024). *Incluyen al guanaco y al pelícano peruano en lista mundial de conservación de especies.* Agraria.Pe. https://agraria.pe/noticias/incluyen-al-guanaco-y-al-pelicano-peruano-en-lista-mundial-d-34870

Aguilar Calla, M., & Torres Zúñiga, D. (2015). *Crianza y mejora genética de las alpacas. Alternativa estratégica para la adaptación al cambio climático.* www.elalvagraf.com

Aguilar, M., Torres, D., Murillo, R., & Zeballos, J. (2014). Buenas prácticas de manejo en la producción de alpacas. *Necesidad estratégica para la adaptación al cambio climático.*

Arzamendia, Y., & Vilá, B. (2012). Effects of capture, shearing, and release on the ecology and behavior of wild vicuñas. *Journal of Wildlife Management, 76*(1), 57–64. https://doi.org/10.1002/jwmg.242

Australian Alpaca Association. (2023). *Sustainability framework.* Australian Alpaca Association.

Cárdenas Minaya, O. E. (2020). *Estrategias de medidas preventivas para enfrentar el cambio climático en la crianza de alpacas | PGC-INIA.* https://pgc-snia.inia.gob.pe:8443/jspui/cris/project/pj00187

CITE. (2013). *Centro de Innovación Tecnológica de los camélidos Sudamericanos* (CITE, Ed.; Primera). Ministerio de Comercio Exterior y Turismo.

Dupuits, E. (2021). *Políticas De Cambio Climático en los Andes.* https://iam-andes.org/wp-content/uploads/2021/11/Politicas_De_Cambio_Climatico_En_Los_Andes.pdf

El Peruano. (2023). Día Nacional de la Alpaca: conoce su importancia para Perú y por qué su lana es apreciada. Noticias. https://www.elperuano.pe/noticia/219550-dia-nacional-de-la-alpaca-conoce-su-importancia-para-peru-y-por-que-su-lana-es-apreciada

El Peruano. (2024). *Oficializan el lanzamiento del "Año Internacional de los Camélidos 2024."* Economía Noticias. https://www.elperuano.pe/noticia/234696-oficializan-el-lanzamiento-del-ano-internacional-de-los-camelidos-2024

FAO. (2005a). *Situación actual de los Camélidos Sudamericanos en Argentina.*

FAO. (2005b). *Situación actual de los Camélidos Sudamericanos en Perú.*

FAO. (2008). *Pro-poor livestock policy and institutional change. Case Studies from South Asia, the Andean Region and West Africa.* https://www.fao.org/documents/card/en?details=2e8db5f3-1a3b-5201-bdd6-747266487bf1/

FAO. (2018). *Soluciones ganaderas para el cambio climático.* http://www.fao.org/gleam/results/en/

FAO. (2024, January 1). *AÑO INTERNACIONAL DE LOS CAMÉLIDOS.* https://www.fao.org/camelids-2024/es

Grupo Inca. (2023). *Why alpaca.* Alpaca Are Gentle Wiyh the Land. https://whyalpaca.com/

Grupo Michell. (2024). *Camélidos Sudamericanos.* https://www.michell.com.pe/michell/es/la-alpaca/camelidos-sudamericanos/

Huanca, T. (1996). *Manual del Alpaquero.*

Huanca, W. (2013). Los desafíos en el manejo reproductivo de los camélidos sudamericanos. *Archivos Latinoamericanos de Producción Animal, 21*(4), 233–236. www.alpa.org.ve/ojs.index/php

Huanca Mamani, T. (2020). *Manual técnico: Producción de alpacas* (Ministerio de Agricultura y Riego, Ed.; Primera). Instituto Nacional de Investigación Agraria. https://hdl.handle.net/20.500.12955/2147

Huanca, T., Naveros, M. L., Gallegos, R., & Mamani, R. H. (2011). Diversidad del color de fibra en alpacas (Vicugna pacos) del Altiplano. *XXII Reunión ALPA, 19*(1), 1–2.

Huanca, T., Naveros, M., Ccopa, J., & Mamani, R. (2018). Problemática y alternativas para mitigar los efectos del cambio climático en la zona alto andina. In *XLI Reunión Científica Anual Asociación Peruana de Producción Animal* (pp. 1–213).

Lanegra, I. (2021). *Política ambiental y del cambio climático.*

MIDAGRI. (2020). *Camélidos sudamericanos.* El Datero Agrario. https://www.midagri.gob.pe/portal/datero/40-sector-agrario/situacion-de-las-actividades-de-crianza-y-producci/298-camelidos-sudamericanos?start=4

MIDAGRI. (2022, August 1). *El 87% de la población mundial de alpacas se encuentra en el Perú.* Plataforma Del Estado Peruano. https://www.gob.pe/institucion/midagri/noticias/636546-midagri-el-87-de-la-poblacion-mundial-de-alpacas-se-encuentra-en-el-peru

MIDAGRI. (2023). Programa de acción de intervención sectorial para el fortalecimiento de las cadenas asociadas a los camélidos sudamericanos 2023-2027 (ANEXO RM No 0276-2023-MIDAGRI). Ministerio de Desarrollo Agrario y Riego.

Ministerio de Desarrollo Agrario y Riego. (2021a). Anuario estadístico: Producción ganadera y avícola 2020. In *Sistema Integrado de Estadística Agraria.* Ministerio de Agricultura y Riego. https://cdn.www.gob.pe/uploads/document/file/2803269/Compendio%20del%20anuario%20%22PRODUCCI%C3%93N%20GANADERA%20Y%20AV%C3%8DCOLA%22%202020.pdf

Ministerio de Desarrollo Agrario y Riego. (2021b). *Perú se mantiene como primer exportador de fibra de alpaca en el mundo por su alta calidad.* Noticias Ministerio de Desarrollo Agrario y Riego. https://www.gob.pe/institucion/midagri/noticias/348402-peru-se-mantiene-como-primer-

Ministerio del Ambiente. (2023). *Sistematización de resultados y lecciones aprendidas del proyecto Ayninacuy.*

Postigo, J. (2009). *Estrategias de adaptación y gestión frente al cambio climático en tres regiones del sur andino peruano.*

Servicio Nacional Forestal y de Fauna Silvestre. (2023). *Perú tiene la mayor población mundial de vicuña – Noticias.* SERFOR Plataforma Del Estado Peruano. https://www.gob.pe/institucion/serfor/noticias/866840-peru-tiene-la-mayor-poblacion-mundial-de-vicuna

United Nations. (2015). *Objetivos y metas de desarrollo sostenible – Desarrollo Sostenible.* Cumbre de Los ODS. https://www.un.org/sustainabledevelopment/es/objetivos-de-desarrollo-sostenible/

Vilá, B., & Arzamendia, Y. (2022a). South American Camelids: Their values and contributions to people. *Sustainability Science, 17*(3), 707–724. https://doi.org/10.1007/s11625-020-00874-y

Vilá, B., & Arzamendia, Y. (2022b). Weaving a vicuña shawl. *Pastoralism, 12*(1), 46. https://doi.org/10.1186/s13570-022-00260-6

Vilá, B., Arzamendia, Y., & Rojo, V. (2020). Environmental education as a means for valuing and conserving camelids and pastoralism in the Argentinean Altiplano of Jujuy. *Mountain Research and Development, 40*(4), D39–D49. https://doi.org/10.1659/MRD-JOURNAL-D-20-00009.1

Vilá, B., Arzamendia, Y., & Rojo, V. (2021). Vicuñas (Vicugna vicugna), Wild Andean Altiplano Camelids: Multiple valuation for their sustainable use and biocultural role in local communities. *Case Studies in the Environment, 4*(1). https://doi.org/10.1525/cse.2020.1232692

Yacobaccio, H. (2009). The Historical Relationship Between People and the Vicuña. In: Gordon, I. J. (eds) The Vicuña. Springer, Boston, MA. https://doi.org/10.1007/978-0-387-09476-2_2

Batik and Textile Supply Chain Management Based on United Nations' Sustainable Development Goals: Indonesian Case Studies

Prima Ditahardiyani, Anastasia Wheni Indrianingsih, Cici Darsih, and Dwi Wiji Lestari

1 Introduction

1.1 Background

Industry is one of the most important economic sectors for a country because it plays a role as a means of investment, labor absorption, creation of higher added value in the commodities produced, fulfillment of domestic product needs, and exports. In achieving this economic development target, there are several challenges for national industry, such as the problem of increasingly limited resource availability, dependence on imported raw materials, and the emergence of industrial waste. Limited resources force industry to implement efficiently at various stages of the production process. Efficient use of raw materials and production processes is expected to produce high-value products without reducing product quality.

Industrial problems related to by-products in the form of industrial waste must also be considered because they have the potential to pollute the environment. Ecosystem damage due to industrial pollution and resource inefficiency can cause

P. Ditahardiyani
Nahdlatul Ulama University of South Kalimantan, Kertak Hanyar, Kabupaten Banjar, Kalimantan Selatan, Indonesia

A. W. Indrianingsih (✉) · C. Darsih
Natural Product Research Group, Research Center for Food Technology and Processing, National Research and Innovation Agency (BRIN), Gading, Gunungkidul, Yogyakarta, Indonesia
e-mail: anas005@brin.go.id

D. W. Lestari
Center for Crafts and Batik, Ministry of Industry, Kota Yogyakarta, Daerah Istimewa Yogyakarta, Indonesia

S. S. Muthu (ed.), *Sustainable Textile and Apparel Chain Management*, Sustainable Textiles: Production, Processing, Manufacturing & Chemistry,
https://doi.org/10.1007/978-3-031-80240-9_6

problems for the environment and communities around the industry. This needs to be a special concern for the government and industry players in preserving the environment so that it does not threaten human health and welfare. At the global level, the demands for the implementation of industrial standards that focus on raw materials, water and energy efficiency, and low-carbon technology are also increasing. The need for green products is also increasing along with the shift in the lifestyle of modern society that pays more attention to environmental aspects. Environmental issues are also one of the barriers to trade for international market penetration. This condition also applies to the textile industry and its derivative products in Indonesia, including batik.

The textile industry has a strong influence on economic development (Pertiwi et al., 2020). Throughout 2019, exports of textiles and garments in Indonesia increased with a value of US$ 13.8 billion (https://www.aseanbriefing.com, October 2023). There are many kinds of traditional cultural heritage in Indonesia; one of them is batik. In Indonesia, batik cloth is mostly produced by female craftsmen in small and medium industries. According to Shaharuddin et al. (2021), several regions in Indonesia that have the most craftsmen include Cirebon, Yogyakarta, Pekalongan, and Solo. Approximately 50.000 batik craftsmen exist in Indonesia. Apart from that, there are also many textile companies that, apart from operating in the field of ordinary commercial fabrics, are also engaged in producing batik cloth. From an economic perspective, the batik and textile industries can increase national income and economic growth. However, from an environmental perspective, these industries cause environmental problems related to the liquid, solid, and gas waste produced by industries. Nindita et al. (2012) reported that each textile industry can produce waste of 202.4 m^3 per month. The industry discharges liquid waste directly into the river, as there are only about 0.6% of industries that have a waste treatment unit (Shofari et al., 2022). This causes environmental damage and endangers human health. The textile and batik industries are paying attention to green technology due to their increased concern for the environment and health. In Indonesia, there are several regulations made by the government to regulate the production industry in Indonesia, such as Local government No. 7 of 2016, Indonesian government No. 27 of 1999, and Ministry of Industry No. 39 of 2019. Those regulations manage the limitation of liquid waste from the batik industry, assessment of environmental impact, and standard of green industry, respectively. With this regulation, the batik and textile industries must fulfill industry requirements following applicable regulations. The aim of these regulations is aligned with 17 goals of SDGs from the UN. Collins et al. (2010) said that there were three pillars of sustainability to achieve the SDGs goals including ecological, social and cultural, and economic. The participation of everyone in the world, including governments, universities, manufacturers, consumers, retailers, and non-government organizations NGOs), and the public is crucial to achieve the SDGs goals. Several studies reported that batik and textile industries in Indonesia are starting to implement green industry and green marketing. For example, Lasem Batik in Central Java carries out the green marketing and brand quality to survive and continue its sustainable production (Wibowo et al., 2023). Fbudi, a fashion brand, used low-impact and sustainable materials such as cotton grown in Flores Maumere and East Nusa Tenggara and natural dyes. Fbudi collaborates with local organizations and local artisans to improve its products. Meanwhile, Sejauh Mata Menadang (Sejauh) uses traditional hand-drawn and hand-stamped batik technique textiles to produce its apparel and homeware products. Sejauh collaborates with local public figures, such as celebrities, photographers, designers, and producers, to promote its products (Tanzil, 2017). Riyanto et al. (2018) revealed that green manufacturing in batik SMEs in Jetis village (Sidoarjo) related to maintaining the waste from the production process, such as wax, wastewater, concentrated wastewater, electrical energy waste, and emissions of Co and CO_2.

Batik is a derivative product of textiles as well as an original Indonesian cultural product. The types, designs, and quality of batik are not static but can be adaptive, developing along with changes in technology, fashion trends, and other cultures that develop in Indonesian society. Each piece of batik cloth can depict the rich culture and history of an area, such as Yogyakarta, Solo, Pekalongan, Cirebon, Lasem, and so on, which are part of the cultural richness of the archipelago.

Based on SNI 0239—2019: Batik—Definition and Terms, batik is a craft art that is created by applying hot wax (batik wax) and using special tools called *canting* to form motifs that have philosophical meaning. Based on the production process, batik is categorized into three types; written batik (*batik tulis*), stamped batik (*batik cap*), and combination batik (*batik kombinasi*). In addition to being a cultural product, batik plays a significant role in the national economy because of its contribution to labor absorption, fulfillment of domestic clothing needs, and as a producer of export foreign exchange with a significant value. The batik industry in Indonesia is spread across 101 centers and involves a workforce of more than 200,000 people.

Batik is a traditional cloth motif and is part of cultural products in Indonesia. Batik has been defined as the process of writing images and decorations on various media using wax as a color-blocking tool (Kamil et al., 2016). The use of wax on batik cloth functions to prevent color absorption during the batik cloth dyeing process. However, ordinary people consider batik only as a cloth that has distinctive patterns and motifs, not based on the technique of making the cloth. Specifically, the term for the tool used to make motifs on batik is called a canting, which is a type of pen made from bamboo and the wax holder is made from brass metal. The word batik itself comes from the proto-Austronesian language "becik," which means making a tattoo, and the Javanese language "amba," or writing and dots (Kamil et al., 2016). In the beginning, the process of dyeing batik cloth used natural dyes, but with advances in technology and the discovery of synthetic dyes, natural dyes were abandoned and replaced with synthetic dyes. However, there are still several batik industries that use natural dyes to preserve culture and reduce pollution caused by the batik dyeing process. Apart from that, batik cloth produced with natural dyes has high economic selling value, distinctive colors, is environmentally friendly, has high artistic value, and has a more exclusive impression (Kamil et al., 2016).

Batik can be divided into three types based on the manufacturing technique, namely "batik tulis" (written batik) or traditional batik, which is made manually by hand. The price of written batik is higher due to the significant amount of care and patience required to make it. The second type of batik is called stamped batik because it uses a stamp or stamp with a batik motif. The stamp used is made of copper. Using this stamp speeds up the batik-making process. The third is printed batik, namely batik printed with modern machines. The process of making printed batik can be done in masse (www.fwd.co.id, 2024).

Batik is part of the textile industry, which provides fabric with various patterns and motifs from all regions in Indonesia (Daud et al., 2023). Batik is made in multiple stages, including making batik motifs, dyeing process using dyes (both synthetic dyes and natural dyes), dye fixing using mordant, wax removal and drying the batik cloth (Ahmad et al., 2002). Batik cloth production usually results in waste that is alkaline, has high COD, has high total suspended solids, and has dark colors (Daud et al., 2022). There are several effects if exposed to synthetic batik dye directly, such as causing symptoms of allergies, skin irritation, and cancer in the long term (Garg et al., 2002). Another thing that worries about the batik industrial process is that there are many home batik industries in Indonesia, some of which dispose of waste from the batik cloth-making process into rivers or waters without going through adequate waste processing. This has triggered researchers to create various treatments to reduce pollution from batik waste, such as through electrochemical processes (Sutisna et al., 2017), biodegration processes (Lun et al., 2022), and the use of plants (Abdulqader et al., 2019).

There are a lot of batik SMEs in Indonesia that use natural dyes since there was an appeal to reduce batik waste, which pollutes the environment and is dangerous for human health. Another reason is that batik cloths produced using natural dyes have a higher economic value. Some natural dye sources that are widely used in Indonesia and sold commercially on the market are tegeran wood (*Cudraina javanensis*), merbau wood (*Intsia bijuga*), mahogany bark (*Swietenia mahagoni*), indigo (*Indigofera tinctoria*), and secang wood (*Caesalpinia sappan*). The use of natural dyes for dyeing batik and textiles in Indonesia continues to be promoted because these dyes are not only sustainable but also nontoxic, non-carcinogenic, biodegradable, and environmentally friendly (Ali et al., 2009; Bechtold et al., 2003). In the process of dyeing batik cloth, mordant also has a big influence on the coloring results. Mordant functions to provide color variations, increase the color strength, and fastness properties (Haji, 2010; Haque et al., 2013). In SME's batik in Indonesia, the mordant used is usually made from chemical compounds such as $FeSO_4$, $CaCO_3$, and $KAl(SO_4)_2 \cdot 12H_2O$. The bio-mordant from natural sources that can be used is soka leaves (*Ixora javanica*) (Darsih et al., 2019); however, this bio-mordant is still rarely used on a large commercial scale.

On the other hand, green supply chain management and sustainability are very important things to do in various industrial processes (Kalkanci & Plambeck, 2020), including the batik and textile industries. The United Nations' sustainable development agenda has 17 sustainable development goals (SDGs) (www.un.org, 2024). Some of the SDGs objectives involve good health, gender equality, zero hunger, no poverty, good education, clean water and sanitation, economic growth, clean energy, industry and innovation, sustainable cities, responsible production, and so on. This has been welcomed by several companies such as Inditex (www.inditex.com, 2024), which regulates its management based on the United Nations SDGs.

The textile industry has supply chain management that is somewhat different from other industries, including a long value chain, intensive work, and generating quite high amounts of waste into the environment (Choi & Cai, 2018). Due to high labor costs, some companies use developing countries to reduce labor costs such as Bangladesh, Indonesia, Vietnam, and Cambodia. However, the lives of workers in the textile industry need to be improved in terms of health, workers' rights, salaries, and safety. Meanwhile, from an environmental perspective, the textile industry usually produces a large amount of pollution to the environment, especially from dyeing fabric using chemicals. This fabric dyeing waste will pollute the environment such as rivers, seas, and waters, if it is not processed properly. On the other hand, the textile industry also consumes large amounts of energy, raw materials, and air for its production process. Meanwhile, fashion trends change very quickly, which also results in high consumer purchases and creates new waste in the textile industry (Dissanayake & Sinha, 2015). Overall, the roadmap in this chapter including research goal and approach (Sect. 1.2), research methodology (Sect. 2), descriptive analysis of searching results (Sect. 3), sustainable supply chain management in textile and batik (Sect. 4), sustainable design of batik and textile (Sect. 5), sustainable process of batik and textile dyeing (Sect. 6), sustainable sourcing of batik and textile (Sect. 7), sustainable production of batik and textile (Sect. 8), sustainable retailing and consumption of batik and textile (Sect. 9), how sustainable batik and textile in Indonesia contributes to SDGs (Sect. 10), managerial recommendations of batik and textile in Indonesia (Sect. 11), and finally, conclusion and future research (Sect. 12).

1.2 The Objective and Approach

This research is intended to explore how the batik and textile industry in Indonesia can be aligned with the Sustainable Development Goals (SDGs) of the United Nations and evaluate whether the batik and textile industry in Indonesia is sustainable in its supply chain management sector. Several previous studies have discussed globally in the field of sustainable fashion supply chain management; however, they have not discussed it comprehensively (Koksal et al., 2017; Choi et al., 2019). This study specifically discusses the batik and textile industry in Indonesia. Thus, this study comprehensively evaluates the batik and textile supply chain management industry in Indonesia based on the United Nations' Sustainable Development Goals. This study uses a systematic literature method to determine prior arts in the batik and textile industry in Indonesia in relation to sustainability and supply chain management.

2 Research Method

This chapter was written using a systematic literature review methodology (Cai & Choi, 2020) to collect and analyze the theme of batik and textile supply chain management in Indonesia. The systematic review method has the advantage of finding developments on related topics and trends for the future without any bias in selecting papers (Denyer & Tranfield, 2009). Therefore, we use this method so that our research objectives can be achieved to increase sustainability in the batik and textile industry in Indonesia and in line with the SDGs from the United Nation. In this method we use several keywords such as "batik industry in Indonesia," "textile industry in Indonesia," "sustainable batik in Indonesia," "sustainable textile in Indonesia," "natural batik dyes in Indonesia," "natural textile dyes in Indonesia," and "supply chain management of textiles in Indonesia." We chose Scopus and Google Scholar as literature search sources and identified papers written in English or Indonesian (with English abstracts). We collect papers that have good academic quality and select papers that suit our research topic. We chose papers with a year range from 2000 to 2024.

3 Analyzing Search Results with a Descriptive Approach

Scopus search results demonstrate that there are around 170 papers discussing "sustainable batik in Indonesia." The number of papers can be increased by adding other keywords. From the collected papers, we selected papers and divided them into several more specific subtopics such as managing the sustainability of supply chains, designing batik and textiles with sustainability in mind, sustainable batik and textile dyeing processes, sustainable sourcing of batik and textiles, sustainable production of batik and textiles, and sustainable retailing and consumption of batik and textiles. From the results of this search, we also identified how sustainable batik and textiles in Indonesia contribute to SDGs and managerial implications and recommendations of batik and textiles in Indonesia.

4 Sustainable Supply Chain Management in Textile and Batik

This chapter explores sustainable supply chains in the textile and batik industries. It begins with a general overview of sustainable supply chain management concept, followed by specific analysis of its application in the textile and batik industries. Supply chain management involves the systematic and strategic cooperation of different business functions in a corporation and across businesses in the supply chain. The objective is to enhance the overall performance of individual companies and the supply chain (Ellram & Cooper, 2014). Many companies have shifted from conventional supply chain management (SCM) to sustainable supply chain management (SSCM) in recent years. This approach focuses on the triple bottom line (TBL) principle, which considers economic, environmental, and social factors. By adopting SSCM, companies address worldwide concerns like global warming and the availability and access to clean and safe water (Rausch-Phan & Siegfried, 2022).

A number of scholars have previously provided definitions of sustainable supply chain management. Gupta and Palsule-Desai (2011) define sustainable supply chain management (SSCM) as emphasizing the importance of managerial practices in reducing environmental impacts during every stage of the product's life cycle. It is crucial for companies to give utmost importance to the environmental impact of their actions when making decisions. This should be considered a significant consideration rather than merely a constraint imposed by regulations or societal pressure. Embracing environmental awareness is crucial, as it should be seen as a long-term commitment rather than a temporary trend. Moreover, it is essential for companies to conduct a comprehensive evaluation of the environmental impacts across the whole value chain, including suppliers, distributors, partners, and customers.

According to Seuring and Müller (2008), sustainable supply chain management refers to managing the flow of material, capital, and information, and fostering inter-company collaboration in the supply chain. This management approach considers sustainable development goals that include economic, environmental, and social dimensions based on customers and stakeholders needs. Another way to define SSCM is the integration of all supply chain players who oversee supply chain operations, assets, and data to optimize financial gain, minimize effects on the environment, and improve social welfare (Tajbakhsh & Hassini, 2015). Based on those definitions, SSCM encompasses three main factors: collaboration among all supply chain partners, equitable consideration of the economic, social, and environmental components of sustainability and a particular emphasis on supply chain stakeholders (Prasad et al., 2020). The SSCM framework focuses on not just the economic aspect but also considers the social and environmental dimensions.

Production costs play a role in the economic aspect of sustainable supply chain management (Bottani et al., 2020), long-term profitability (Gómez-Luciano et al., 2018), green purchasing performance (Seman, 2012), waste reduction (Özceylan et al., 2017), customer satisfaction (Soykoth et al., 2022), competitive advantage (Luthra et al., 2015), and level of inventory (Swarnakar et al., 2021). The social dimension of sustainable supply chain management studies occupational health and safety (Ahi & Searcy, 2015), corporate social image (Xia & Li-Ping Tang, 2011), green image and product image (Seman, 2012), workforce absorption capacity (Ahi & Searcy, 2015), corporate social responsibility (Hsueh, 2015), and employee welfare (Gopal & Thakkar, 2016).

Various research studies on sustainable supply chain management (SSCM) employ different indicators to assess environmental aspects. These indicators include green design level, green purchasing level (Seman, 2012), decrease in the utilization of raw materials (Qi et al., 2011), and reducing emissions (Nepstad et al., 2013), low energy and hazardous material consumption (Gray & Charter, 2007).

Table 1 Dimensions of sustainable supply chain management

Economic dimension	Social dimension	Environmental dimension
Production costs Long-term profitability Green purchasing performance Waste reduction Customer satisfaction Competitive advantage Inventory levels	Occupational safety and health Corporate social responsibility Workforce absorption capacity Corporate social image Employee welfare Green image	Green design level Optimizing the use of raw materials Green purchasing level Minimizing emissions and deforestation Energy consumption Usage of toxic/hazardous/harmful materials

Table 1 summarizes this review of SSCM dimensions, which can be implemented to evaluate sustainable supply chain management implementation.

Successful application of sustainable supply chain management requires the presence of drivers. Diabat et al. (2014) identified the elements that contribute to the successful implementation of sustainable supply chain management (SSCM), which contain internal drivers (from within the organization) and external drivers (from external sources). Internal drivers typically include managerial mindset, employee expectations, corporate culture, and social development initiatives. In contrast, external factors include customer demand, pressure from investors, community groups, society, competitors, and the need to comply with regulations.

4.1 Sustainable Supply Chain Management in Textile Industry

The textile industry is highly vulnerable to sustainability-related issues. The industry is distinguished by an exceedingly competitive atmosphere. To achieve success, the textile industry must show adaptability to accommodate evolving consumer requirements, while simultaneously ensuring cost efficiency to offer affordable prices. Textile companies generally depend on just-in-time production and fast fashion as their main supply chain strategy. It encourages the development of partnerships with low-cost suppliers that comply with less stringent social and environmental regulations to maintain a competitive advantage.

The textile supply chain is renowned for its labor-intensive nature, lengthy value chain, and substantial environmental impact, in comparison to supply chains in other industry (Cai & Choi, 2020). From an environmental standpoint, the textile and apparel industry significantly contributes to environmental pollution. The textile production process generates waste that contains hazardous compounds, particularly during the dyeing process (Siddique et al., 2017). If not well managed, the rivers and streams can become severely contaminated. Unfortunately, most of these factories do not prioritize investing in clean technologies for wastewater treatment. From an economic and social standpoint, textile production capitalizes on the cost advantages of cheap labor in developing countries. This brought attention to the public regarding the rights, health and safety, and payment of workers. In addition, the prevalence of fast fashion and rapidly changing fashion trends has resulted in expanded consumer spending, a decrease in the lifespan of clothing, and an increase of waste produced (Dissanayake & Sinha, 2015). Therefore, the textile industry has a significant challenge in attaining the sustainable economic, social, and environmental goals.

There are internal and external barriers to implementing sustainable supply chain management. Initially, internal barriers are characterized by the following: insufficient resources, limited awareness, commitment, and support from the top management level to prioritize sustainability issues, a lack of knowledge, low management skills, and inadequate training (Brammer & Walker, 2011). Second, regulatory issues are the focus of external barriers. Regulation, in this case, merely calls for meeting the bare minimum, which stifles creative, sustainable supply chain management procedures or prevents proactive conduct. Moreover, an external barrier to sustainable supply chain management includes the buyer-supplier relationship, price consciousness in purchasing decisions, a lack of demand for sustainable products, and competitive pressure (Oelze, 2017).

The sustainability dimension based on the "3BL" framework is important in the textile sector. Chi (2011) examines the strategic management aspect of enhancing the sustainability of supply chain management. The study highlights that corporate social responsibility (CSR) is crucial in achieving sustainability objectives in global fashion supply chains. Oelze (2017) examines the factors that support and hinder sustainable supply chain management and concludes that partnerships, like cross-industrial collaboration and cooperation among supply chain participants, can enhance sustainable supply chain adoption. The operations management examines the approach of textile and fashion firms towards sustainability development (Na & Na, 2015). They found that environmentally friendly materials, recycling products, and green marketing can all contribute to sustainable development. In addition, according to Atilgan (2007), eco-labeling is essential to addressing issues related to human health and the environment in developing the sustainable textile sector.

4.2 *Sustainable Supply Chain Management in Indonesian Textile Industry*

There are several obstacles that Indonesian textile industry must overcome to enhance supply chain sustainability. The lack of sustainability pressure from investors and consumers, as well as the lack of support from the government, has resulted in most textile industries still needing to implement sustainable supply chain management (Gunawan et al., 2022). With external pressure and especially with government support, it is easier for textile industry players in Indonesia to prioritize sustainable supply chain applications, as these applications require costly resources. Other barriers were scarce local raw material resources and insufficient knowledge and education. Indonesia is significantly falling behind its competitors in the level of education and when considering sustainability. There is a perceived need for national legislation to incentivize firms to take action and govern and control the industry.

The supply chain in Indonesian textiles faces challenges in all three dimensions of sustainability. From the environmental dimension, the textile and apparel industry in Indonesia should apply green practices, for example, green design, green distribution, eco-design, green marketing, and green manufacturing to improve the sustainability performance of supply chain management (Srisawat & Srisawat, 2020). Considering the social aspect, Noerhartati et al. (2020) suggest that firm increase management commitment and foster an organizational culture that emphasizes the significance of green industrial practices and allocates adequate time for implementation. From an economic perspective, Wongwilai et al. (2022) recommended that textile industry need to improve their ability generating innovative ideas to sustain business functions, like promoting the latest product innovations and using the latest technology in both production processes and supply chain management.

Evaluating performance of sustainable supply chains to enhance organizations' competitiveness by implementing sustainable practices is imperative. Nisrina and Ardi (2022) devised a methodology to assess the sustainable supply chain performance of the textile industry in Indonesia by considering three dimensions: economic, social, and environmental. Six important indicators comprise the economic dimension. The first indicator is related to the total cost incurred along the supply chain from raw material procurement until the product reaches the consumer, such as raw material cost, production cost, transportation cost, labor cost, inventory cost, and others. The second indicator relates to the costs incurred for employee training and education, while the third indicator is the level of product quality. The fourth and fifth indicators are the time it takes for products to reach consumers and the inventory levels of raw material, by product and finished goods, respectively. The final indicator is the level of customer satisfaction with products and services.

The environmental dimension is divided into 14 key indicators (Nisrina & Ardi, 2022). Six indicators relate to the production inputs used, namely energy usage, materials usage, water usage, organic material usage, use of green technology, and the usage of hazardous chemicals in the production process. The following five indicators relate to the quality of the surrounding environment due to production, distribution, and transportation along the supply chain, consisting of water quality and water pollution levels, air quality and air pollution levels, soil quality and soil pollution levels, the amount of CO2 gas emissions generated, and the amount of noise generated. Other environmental indicators include a waste management system.

The social dimension is made up of 11 key indicators. These indicators include the number of workplace accidents, the absence of underage workers, the availability of education and training for workers to improve skills and competencies, decent wages, worker welfare, the lack of discrimination and intimidation in the workplace, equal opportunities for all workers, protection from sexual violence, and the empowerment of local communities (Nisrina & Ardi, 2022).

According to Sarasi et al. (2023), there are several factors for sustainable supply chain management performance in the Indonesian textile industry. These are corporate entrepreneurship, organizational culture, competitive action, innovative ideas and processes, PLC assessment, and business sustainability typology. Furthermore, implementing eco-design, green trust, and innovative supply chains are other substantial factors in gaining sustainability. Effective sustainable supply chain management performance necessitates both internal and external drivers. External drivers can be classified into three categories: market pressure, societal pressure, and regulatory pressure (Sarasi et al., 2023). Market pressure refers to a variety of factors that influence the company's operations and performance, including pressure from competitors, shareholders and investors, institutional pressure, pressure from suppliers, pressure from customers, and pressure related to reputation and image. Regulatory pressure, on the other hand, refers to the influence of government regulations, regional and international bodies such as the European Union, professional or trade associations, and various certifications that increase competitive advantage and economic benefits. On the other hand, societal pressure is an external force that influences the actions and decisions of companies, including pressure from non-governmental organizations (NGOs), media and press pressure, public pressure, consumer organization pressure, and public welfare pressure.

Corporate strategy, organizational culture, organizational resources, and organizational characteristics are the four groups of internal drivers (Sarasi et al., 2023). The organization's code of conduct, business ethic organization, level of innovation, information dissemination, attention to employee health and safety, and organizational responsibility to society and others reflect its organizational culture. Organizational strategy includes top management commitment, organizational strategy related to cost policy, and strategy related to achieving organizational performance. In contrast, organizational resources consist of the use of organizational resources, skills and abilities of human resources, technology and equipment owned by the organization, and training and education provided by the organization. Organizational characteristics consist of the position of the organization in the supply chain, the geographic location of the organization, the size of the organization, and the level of sustainability of the organization.

4.3 Sustainable Supply Chain Management in Indonesian Batik Industry

The Indonesian economy is greatly influenced by the batik industry. However, the batik industry causes environmental damage. The batik production process, which still uses hazardous materials and inefficient resources, causes water and air pollution (Handayani et al., 2018). As a result, sustainability is critical in the batik industry.

The sustainability of the supply chain in the batik industry is generally linked to the environmental dimension. So far, Indonesia's batik industry still needs help achieving water, energy, and resource efficiency (Yaacob & Mat Zain, 2016). Moreover, batik production wastewater contains harmful elements, heavy metals, organic compounds, and other materials that cannot be biodegraded. These pollutants are frequently released into rivers without enough processing (Handayani et al., 2018). Discharging toxic wastewater into rivers can pose a significant health hazard to people, fauna, and flora living near the river (Yaacob & Mat Zain, 2016).

SMEs dominate the batik industry in Indonesia. One of the characteristics of batik SMEs is a need for more understanding of the risks associated with the use of dangerous and poisonous substances in manufacturing process especially in the dyeing process. Using these materials results in environmental pollution, which also harms artisans. Due to a lack of knowledge, batik SMEs have yet to utilize environmentally friendly natural dyes. In addition, batik entrepreneurs do not have sufficient knowledge about the advantages and disadvantages of implementing sustainable practices. This lack of knowledge is particularly prevalent among women with limited educational opportunities and access to business networks. These constraints prevent women from developing the necessary skills and capabilities to implement sustainability in the batik industry (Yoshanti & Dowaki, 2017).

In terms of energy usage, previous studies have indicated that batik small and medium enterprises (SMEs) frequently rely on nonrenewable energy sources, such as kerosene, throughout their manufacturing processes. Excessive energy consumption contributes to the degradation of the ozone layer and the harmful effects of hazardous chemicals on terrestrial ecosystems (Yoshanti & Dowaki, 2017). Moreover, Batik small and medium enterprises (SMEs) inadequately utilize resources, which is a concern too. Prior research has indicated that a significant amount of waste in the batik manufacturing process originates from faulty items created during pattern creation, waxing, and dyeing and from excessive production. Waste is also generated during sewing and packing operations (Handayani et al., 2018). It is imperative for this industry to enhance their production efficiency, prioritize environmental protection, and actively contribute to society. SMEs can enhance sustainability by improving ecological sustainability.

From an economic dimension, variables like changes in the market price of raw materials or the devaluation of the local currency might impact the sustainability of batik products. The primary components in batik manufacturing include textiles, wax, chemicals, and dyes. Unfortunately, most batik materials had to be imported from China, India, America, and Germany. As a result, Batik SMEs have a restricted ability to choose the prices of their products. Consumer awareness of sustainability also played a role in influencing the outcome. Most batik consumers are older women with low education levels who need help understanding the importance of sustainability. As a result, the market share for sustainable batik still needs to grow.

Regarding the social dimension, ergonomic research has demonstrated that musculoskeletal diseases (MSDs) are experienced by 60.2% of batik laborers, resulting in discomfort in the ankle, lower back, and shoulder. Previous research (Siswiyanti et al., 2019) showed a significant prevalence of musculoskeletal disorders (MSDs) among batik workers. The research found that the dyeing process, which requires workers to stoop and crouch, has contributed to musculoskeletal issues in several areas of their bodies. This poses a considerable socioeconomic burden and greatly affects the workers' health and performance quality.

The sustainability criteria of batik SMEs were evaluated using a comprehensive three-dimensional framework that encompasses the economic, social, and environmental aspects (Hasibuan & Hidayati, 2018). Economic dimensions include market growth, market share, profit, enhanced consumer benefits, extended business development, marketing collaboration, and impact on the macroeconomy. The environmental dimension encompasses self-life optimization, transportation, and distribution efficiency, resource conservation, waste minimization, toxicity concerns, and the promotion of reuse and recycling. On the other hand, the social dimension encompasses social responsibility, occupational health and safety, work environment conditions, employment, the connection between justice and employment, the preservation of regional traditions, and the protection of property rights, brands, and patents.

Previous study (Hasibuan & Hidayati, 2018) states that batik industry status in Indonesia is moderately sustainable. One aspect contributing to Batik SMEs' sustainability is a firm commitment from senior management to adopting a cleaner manufacturing, innovation, and creativity strategy. Successful cleaner manufacturing technologies mostly focus on substituting hazardous dye ingredients with nontoxic alternatives, reducing the energy usage, and promoting the recycling of waste.

For improving ecological sustainability, Batik SMEs can practice ecological entrepreneurship (ecopreneurship). Ecological entrepreneurship, also known as ecopreneurship, in the batik industry involves several techniques. These strategies include the utilization of natural dyes, improving resource efficiency by conserving water, textiles, wax, and dyes (Handayani et al., 2018), and utilizing alternative energy sources (Yoshanti & Dowaki, 2017). Transitioning to natural dyes offers environmental protection and enhances safety for both artisans and consumers, particularly in terms of skin health. In relation to resource efficiency, previous research has demonstrated that by minimizing waste and implementing recycling practices, an SME has the potential to cut production costs and result in an increased efficiency rate and positively impact the local economy (Rizos et al., 2016). It is commonly recommended that SMEs utilize local resources and recycle raw materials.

Moreover, formal institutions can facilitate enhancing ecological sustainability through policy initiatives. Policy interventions, such as subsidies and regulatory adaptation, can assist newly established sustainability-focused companies in their first phases, aiding their survival. Sustainability-oriented policies can provide intangible advantages, including improving the image and reputation of the community, attracting skilled and dedicated staff, and preventing non-compliance with socio-ecological sustainability standards (Maniu et al., 2021). Furthermore, government-led knowledge transfer projects have the potential to enhance environmental standards and the competitiveness of new enterprises greatly.

The government is instrumental in facilitating and supporting female entrepreneurs in the process of establishing more sustainable enterprises. This statement challenges the idea that women should embrace entrepreneurial empowerment based on their individual aspirations and principles, including those related to sustainability. Therefore, it is crucial to revise sustainable entrepreneurship programs in order to empower women-led small and medium enterprises (SMEs). This may be accomplished by providing women with more opportunity and flexibility to actively engage in the initiatives aimed at enhancing the sustainability of their enterprises. It is important to take into account various age groups and levels of education.

To improve economic sustainability, Batik SMEs should establish relationships with suppliers of raw materials and finance companies to guarantee a consistent supply chain. Nearly all SMEs are defined by their small scale, limited financial resources, restricted access to external funding, and increased vulnerability to bankruptcy. Gunawan et al. (2022) propose strategies to enhance economic sustainability in the SME sector. They suggest the development of machines to improve production efficiency while preserving the authenticity of batik and recommend implementing a digital system to archive batik-related data and prevent data loss. They also conduct workshops to assist SMEs in utilizing the system effectively.

5 Sustainable Design of Batik and Textile

The textile industry is one of the manufacturing industries that contributes greatly to the Indonesian economy because it produces products with high export value, is labor-intensive, and is capital-intensive. The textile industry consists of the spinning, weaving, and textile refinement industries, which produce fiber, thread, cloth, apparel, and their derivative products. Batik is a textile derivative product that has long been developed in Indonesia. The batik industry is one of the textile derivative industries that experienced rapid development and has encouraged mass batik production in Indonesia.

However, along with its development and contribution to the economy, the textile and batik industry in Indonesia still leaves environmental problems, such as raw material and energy inefficiency and its status as the biggest cause of pollution because the waste contains chemicals, including dyes, in high levels (Kanchana et al., 2013). Wastewater is still a problem in almost all textile companies and batik centers in Indonesia. Liquid waste from the textile and batik industry contains chemicals such as waste from dyeing processes and wax removal processes.

Solid waste, such as batik wax, is produced during the batik production process. The remaining wax from the batik process will solidify when the temperature cools, and if not handled properly, the used wax solids will become waste, which will certainly damage the environment. Another solid waste is scrap fabric waste. This waste is produced due to inaccuracies in measuring and cutting fabric in the production process of clothing, accessories, and other products. Fabric waste can also be formed due to improper storage and handling methods. Fabric materials that

Table 2 Best practice sustainable design of batik and textile

No	Aspects	Best practices
1	Renewable	Fabrics from natural fibers such as cotton and rayon as batik media Stearin wax made from palm oil as a substitute for fossil-based paraffin wax Natural dyes are an alternative to chemical dyes
2	Efficient	Effective utilization of raw materials in accordance with Green Batik Industry Standards.
3	Safe or nontoxic	Doesn't use dangerous chemicals, such as dyes which containing azo.
4	Cyclic	Recycle wax waste for reuse
5	Social	Making products from waste materials through recycling and upcycling from waste fabric scraps into aesthetic products with economic value.

are stored for too long can become damaged and dirty, so they cannot be used. For the batik and textile industry in Indonesia to continue to be competitive and sustainable, this industry must pay attention to the principles of sustainability through a sustainable design approach (Tischner, 2011).

Sustainable design is really needed by society at large because of the emergence of pressures on the environment, such as environmental pollution, increasing population and human needs, the greenhouse effect, global warming, and so on. Sustainable development is the right step in dealing with the environmental crisis that occurs in almost all countries because of development (Arumsari, 2020). Sustainable design that can be applied to the textile and batik industry is an environmental and ecological principle that focuses on material and energy efficiency, the use of renewable energy, and the application of reduce, reuse, and recycle in the production process to minimize activities or materials that damage the environment (Fadhilah & Widiawati, 2021). Best practice sustainable design of batik and textile is presented in Table 2.

There are several studies related to sustainable production in the batik industry in the form of eco-efficiency studies of batik production (Nindita et al., 2012; Sari et al., 2012). Other research focuses on studying opportunities for implementing cleaner production in the batik industry (Nurdalia, 2006) and developing sustainable development models in the batik industry (Fadhilah, & Widiawati, 2021). The results of this research show that the batik industry in Indonesia is generally still less efficient in using materials such as fabric, dyes, and wax. Most batik industries, which are small or home-based industries, have not yet handled and processed waste with waste processing installations (IPAL) because funding is still considered expensive (Indrayani, 2019).

The sustainable design approach includes several aspects, including renewable, efficient, safe or nontoxic, cyclic, and social (Tischner, 2011).

(a) *Renewable*

Products are made using raw materials and energy from renewable sources. Efficiency in manufacturing design products can be achieved with several strategies, such as dematerialization of materials and energy, optimizing product usability and multifunctionality, and increasing product durability and lifespan.

(b) *Safe or nontoxic*

Product does not contain or produce substances harmful to life and the environment.

(c) *Cyclic*

Products are in a cyclic life cycle, both in terms of materials and manufacturing techniques, for example, trying to recycle waste into new materials that can be reused with good quality for various other uses.

(d) *Social*

Products made can be used as social activities to improve the economic capacity of the community as well as environmentally friendly campaign media.

6 Sustainable Process of Batik and Textile Dyeing

The dyeing process for coloring batik and textiles is a critical factor in batik industry. Most batik and textile industries use synthetic or chemical fabric dyes. This is executed because synthetic dyes for batik and textiles are very easy to obtain, have a large supply and variation of colors, do not fade easily, their availability is guaranteed, and the prices are quite affordable. However, artificial colors in batik and textiles have the potential to negatively impact both human health and the environment because heavy metals such as tin (Sn), chromium (Cr), zinc (Zn), lead (Pb), and copper (Cu) are commonly found in synthetic dyes. Long-term accumulation of heavy metals in the human body can cause chronic diseases and carcinogenicity. The use of natural dyes that are more sustainable and environmentally friendly is being urged due to this. Even though there are still many limitations found in natural dyes, such as fading easily, not many colors available, and the dyeing process being more complicated, considering the risk of synthetic dyes, as well as their long-term effects on humans and the environment, the use of natural dyes in batik and textiles is a better step.

The dyeing process has a negative impact on the environment, including water pollution. Synthetic chemicals remaining from the dyeing process can cause disease for batik industry workers and the surrounding community if not handled properly (Cai & Choi, 2020). Several factors influence the sustainability of the dyeing process, for example, the choice of batik and textile dyes whether to use natural or synthetic dyes and the technical implementation of dyeing. A study shows that the dyeing method using spin-dyeing in the textile industry will reduce the use of energy, water, and carbon emissions in order to minimize negative environmental effects compared to conventional dyeing processes (Terinte et al., 2014). Another study shows that a new dyeing method using enzyme and plasma pretreatment increases sustainability in textile dyeing and reduces the use of water, energy, and raw materials (Parisi et al., 2015).

In Indonesia, there are many natural sources from nature that can be used as dyes in batik production. These materials come from plant parts in the form of stems, flowers, fruit, and roots (Indrianingsih & Darsih, 2013). Parts of the banana tree (*Musa paradisiaca*) that can be used as natural dyes include the trunk, petiole, and fruit stalk, which can produce a brown color. The cinnamon plant (*Cinnamomum verum*) can also produce a brown color from the stem. The roots of the noni plant (*Morinda citrifolia* Linn) can produce natural dyes in red, yellow, and brown. The leaves of the tarum plant (*Indigofera tinctoria* Linn) can produce a blue color. Parts of the plant in the form of rind from mundu (*Garcinia dulcis*) can produce green dye. Meanwhile, the skin of the mangosteen fruit (*Garcina mangostana* L) can produce red and violet colors. The leaves of the betel plant (*Piper betle*) can also provide a natural brown color. The fruit of the areca plant (*Areca catechu*) can give a red color. Meanwhile, the leaves and young branches of the gambier plant (*Uncaria gambir*) can produce red and black colors. The leaves of the mango plant (*Mangifera indica*) can provide a natural green dye, the same as the leaves of the sengon plant (*Albizia chinensis*) which also produces a green color. Meanwhile, natural brown dye can be produced from the stems of the coastal merbau plant (*Intsia bijuga*), the stems of the mahogany plant (*Switemia mahogany*), and tea leaves (*Camellia sinensis*). The leaves and wood of the teak plant (*Tectona grandis*) can produce a natural brown dye, while pineapple (*Ananas comosus*) can produce a red color. The roots and bark of soga (*Peltophorum pterocarpum*) produce yellow, as do the rhizomes of turmeric (*Curcuma longa*), which produce yellow dye. The seeds of the keeling crimson plant (*Bixa orellana*) can also produce orange and red dye. The granule fruit part of the *Mallotus philippensis* plant also produces an orange color. Natural red dye can also be produced from the bark of the dragon fruit plant (*Hylocereus polyrhizus*), the bark of the sappan plant (*Biancaea sappan*), the wood of the *Aporosa frutescens*, the leaves, and young branches of the magenta plant (*Peristrophe bivalvis*). Meanwhile, natural brown dye can be achieved from the leaves and seeds of a combination of the *Uncaria gambir* plant, *Piper betle* leaves, and *Areca catechu* seeds. Apart from that, the brown color can also be achieved from the bark of the soga plant (*Peltophorum pterocarpum*), and wood from the acacia plant (*Acacia catechu*). Green natural dye can be obtained from suji leaves (*Dracaena angustifolia*), while blue natural dye can be obtained from seaweed and leaves from tarum plant (*Marsdenia tinctoria*). Natural dyes that produce yellow include flowers from the palasa plant (*Butea monosperma*), wood from the tegeran plant (*Maclura cochinchinensis*) and bark from the jirak plant (*Symplocos fasciculata*). The natural black dye can be produced from the bark, leaves, root, and fruit of the ketapang plant (*Terminalia catappa*).

We have previously experimented with mangosteen peel waste (*Garcinia mangostana*) as a natural dye for batik (Indrianingsih et al., 2021). Water that contains citric acid or ethanol can be used to carry out the extraction process. The dyeing process uses mangosteen peel extract followed by the use of mordant. The mordant could be a chemical mordant such as calcium hydroxide ($Ca(OH)_2$), ferrous sulphate ($FeSO_4{\cdot}7H_2O$), or alum ($KAl(SO_4)_2{\cdot}12H_2O$). The natural mordant using *Ixora javanica* leaves. The results obtained show that dyeing batik cloth with mangosteen peel extract can produce quite good colors and produce functional cloth that has antibacterial properties. The functional value of antibacterial properties is a distinct advantage of utilizing mangosteen peel waste. The results of this antibacterial fabric can be used in the medical and pharmaceutical industries, such as doctor and nurse uniforms, hospital curtains, athlete uniforms, hospital beds and linens, etc.

Research on the use of Tegeran (*Cudrania javanensis*) extract has also been carried out in our previous study for dyeing batik cloth (Darsih et al., 2019). Apart from natural dyes, bio-mordants from nature are also used, namely soka leaves (*Ixora javanica*). The use of biomordant from soka leaves is known to have comparable results to the commonly used alum mordant. The results obtained showed that soka leaf extract produced quite good coloring and good fastness. The use of natural dyes and mordants can ensure the sustainability of the batik dyeing process so that it will reduce chemical waste produced from the batik dyeing process. The health of batik workers is also better guaranteed by using natural dyes and natural bio-mordants.

SME's batik in the Temanggung area, Central Java, Indonesia, also uses natural indigo blue dye, which comes from extracts from the *Strobilanthes cusia* plant (Kurniawan et al., 2021). This plant is utilized to produce natural indigo dye because the yield produced is greater than other plants. From May to August every year, *S. cusia* is harvested to produce natural dye. The process of making *S. cusia* extract for natural dye begins by soaking the *S. cusia* in water for 2 days. The resulting liquid will turn greenish yellow, and then it will be filtered to separate it from the biomass. The next step is the aeration process to increase the oxygen content, thereby creating indigo dye. The pH should be adjusted to around 11–12 in the final stage of creating natural dyes. The indigo dye that has been formed is then stored in paste form at room temperature. Apart from *S. cusia*, another plant commonly used to produce indigo dye paste is the *Indigofera tinctoria* plant (Ariyanti & Asbur, 2018).

Apart from using natural dyes to increase sustainability in the batik dyeing process, other efforts are also being made to reduce the batik waste produced using natural materials for the bioremediation process. A previous study used the *Ganoderma lucidum* fungus to reduce batik waste in batik SMEs located in the Petir area, Rongkop, Gunungkidul, Yogyakarta, Indonesia (Pratiwi et al., 2017). The bioremediation process was studied using incubation time and dye concentration parameters. *G. lucidum* fungus was grown on batik waste with various concentrations at a temperature of 25 °C and pH 5–6. The results obtained indicate a decolorization process of batik waste with increasing incubation time. This indicates that the process of reducing waste and purifying waste can be done with natural ingredients, namely the *G. lucidum* fungus.

7 Sustainable Sourcing of Batik and Textile

One of the strategies of the Indonesian government in realizing SGDs in the industrial sector is to prepare Green Industry Standards (GIS) and organize green industry certification, which is regulated in Industrial Law No.3/2014. GIS was prepared as a government tool to develop industry and at the same time can be a guide for industry. Specifically, GIS for batik and textiles has also been prepared by the Ministry of Industry as stated in Minister of Industry Regulation No.10/2023 concerning Batik SIH and Minister of Industry Regulation No.40/2022 concerning SIH for the textile industry. Some of the technical requirements regulated in the green industry standards are related to efficiency in the use of materials and energy, the use of low-carbon technology, and aspects of waste management.

The materials needed in the batik production process are fabrics, batik wax, and dyes. Basically, not all fabrics can be used to make batik. The fabric used for batik must meet technical requirements, including not being damaged by the batik process and being able to be colored at room temperature. Fabrics produced from natural fibers, such as cellulose fibers and protein fibers, can meet these requirements (Suparno, 2020; Fika Handayani, 2009). Cotton and silk are the natural fibers most widely used in making batik today. Cotton fabric is a type of cellulose fiber made from cotton, while silk is a type of protein fiber produced from silkworm cocoons.

Indonesia needs cotton fiber in very large quantities, and most of it is met from imports. Therefore, efforts are needed to utilize natural fibers other than cotton as alternative raw materials. This aims to anticipate the possibility of a shortage of cotton fiber in the country. Natural fiber plants that have the opportunity to be used as alternative raw materials or supplements for cotton fiber are hemp, abaca, jute, kenaf, sisal, linum or flax, and kapok (Puslitbangbun, 2019). In the future, research regarding the use of alternative natural fiber fabrics to be used as batik raw materials needs to be carried out.

Batik wax, or *malam*, is one of the important ingredients in the batik-making process because it functions as a color-blocking agent (Atika & Haerudin, 2013). In general, the allocation of production costs for purchasing batik wax reaches approximately ±70% of the cost of batik material requirements. The raw materials used in making batik wax consist of seven types: damar (resin of the Shorea Sp. tree), gondorukem resin (residue of distillation of merkusii pine resin), kote (bees wax), paraffin (a by-product of oil processing raw), microwax kendal (animal fat), and used wax (Susanto, 1980).

Batik production in Indonesia still uses wax from paraffin formulations. As is known, paraffin comes from petroleum and is predicted to slowly run out because it is nonrenewable energy. This could threaten the continuity of textile and batik industry. Moreover, most of the paraffin is still obtained by import. Research collaboration between the Center for Craft and Batik, Ministry of Industry, and the National Research and Innovation Agency (BRIN) has produced a batik wax formula made from renewable raw materials. Stearin, which is a solid fraction of processed palm oil, can replace the function of paraffin obtained from petroleum.

Dyes are an important raw material in batik production because the color of the product is one aspect of consumer criteria in choosing batik. In addition to being influenced by the level of complexity of the motif, the beauty of batik colors can add to the artistic value of the product. Therefore, the selection of the type of dye, the quality of the dye, and the right coloring technique are some of the determinants of the quality of batik products. There are two types of dyes used in textiles: natural dyes and synthetic dyes (Widyasti et al., 2017). Natural dyes are dyes that come from nature, usually from plant parts such as roots, tubers, stems, twigs, leaves, bark, wood, seeds, and flowers. While synthetic dyes are artificial or synthetic dyes made through chemical reactions. Several types of synthetic dyes for batik coloring are remazole, indigosol, and napthol.

The use of synthetic dyes for textile products has several advantages, including more affordable prices, more stable color results, a more practical coloring process, and easy to get in stores. However, the use of this type of dye has a negative impact on the environment and health if waste management is not carried out properly (Subagyo & Soelistyowati, 2023; Pujilestari, 2015; Kusumawati et al., 2019). Although not all types of synthetic dyes are dangerous, some synthetic dyes are reported to be hazardous materials and can cause environmental pollution because they can be degraded into compounds that are carcinogenic and toxic (Widjajanti et al., 2011; Kant, 2012). Therefore, when using synthetic dyes, proper and good processing of the waste produced from the dyeing process is required.

Nowadays, natural dyes are starting to be widely used again for several environmental reasons, such as their waste is easily biodegradable, they come from renewable plant sources, and they have a lower negative impact on human health and the environment (Bechtold et al., 2003; Kusumawati et al., 2021; Yernisa, et al., 2013; Ali et al., 2016). The batik industry in Indonesia has used many natural dyes in its production process, such as tingi tree bark (Ceriops tagal), tegeran wood (Cudraina javanensis), coconut fiber (Cocos nucifera), mahogany bark (Swietenia mahagoni), and indigo (Indigofera tinctoria) (Darsih et al. 2019; Susanto, 1980).

Apart from the dyes, there are several additional chemicals needed in the dyeing process using synthetic dyes, such as sodium carbonate (Na_2CO_3), sodium metasilicate (Na_2SiO_3), and sodium hydroxide (NaOH). Meanwhile, dyeing batik using natural dyes also requires mordant. Mordant is a chemical that forms complex compounds with dyes and bonds strongly with fabric fibers. The use of mordant in the natural batik coloring process has been widely researched and reported. The results of the study showed that the use of mordant in the dyeing process can provide more color variations and can improve the quality of the dyeing results, including the color fastness of the dyed fabric (Atika & Salma, 2017; Lestari et al., 2020). Although there are several types of mordants, the majority of the batik industry only uses metallic mordants, there were lime ($CaCO_3$), ferrous sulphate ($FeSO_4$), and alum ($Al(SO_4)_2 \cdot 12H_2O$). Other types of mordants are tannin mordants, such as *soka* leaves, *Symplocos fasciculata* leaves, *Psidium guajava* leaves, and *Peltophorum pterocarpum bark* (Lestari et al., 2020; Darsih et al., 2019).

The use of natural-based materials such as natural fibers, natural dyes, and natural mordants in batik and textile production is considered more environmentally friendly because it is classified as a renewable material. These materials can also be easily obtained throughout Indonesia. Moreover, Indonesia is a country with abundant natural resources, so it has great potential in providing raw materials from nature (Rosyida & Zulfiya, 2013). In addition to reducing dependence on imported raw materials, efforts to use natural materials are also a real manifestation of advancing local industry while saving the environment.

8 Sustainable Production of Batik and Textile

Sustainable production of batik and textiles in Indonesia was related to many factors such as material used in industry, energy, waste produced in the production process, labor's rights, etc. The regulation from the Ministry of Industry No. 39 of 2019 is expected to guide companies in carrying out efficient and environmentally friendly production processes. And one of the industries that must comply with this rule is the textile and batik industry. Some studies reported that SMEs batik implemented green industry in its production. Currently, the batik industry has begun to switch to use wax made from natural palm oil, which replaces wax from fossil fuels (https://www.news.palmoilmagazine.com). In Roundtable on Sustainable Palm Oil (RSPO) 2023, one of the batik producers, Laweyan Solo, Central Java, marketed batik that is used palm wax (https://www.palmoilmagazine.com/rspo/2023/11/25/laweyan-sustainable-palm-batik-started-to-be-marketed-in-rspo-rt2023/). Palm wax is considered more environmentally friendly, and the raw materials for making it are abundant in Indonesia. Shofari et al. reported that the factor of green productivity was dominated by waste treatment. The authors calculated the Environmental Performance Indicator (EPI) index in Batik Ayu Arimbi, Indonesia (Shofari et al., 2022). The EPI index can show how the environmental performance of the batik industry. The industrial sustainable development was promoted by clean manufacturing strategy implementation (Ajizah et al., 2021). Technical, economic, and environmental aspects were used to assess and determine cleaner production in Mbah Guru factory in East Java. The authors found that the reuse of natural dyes is the best method for implementing cleaning practices in the batik industry. Kusumastuti et al. (2023) surveyed the sustainable production practices of 209 batik producers by completing paper questionnaires and using electronic forms. The authors found that this study is an important step towards sustainable production and supports the strengthening of the concept of natural batik dye production. Syahputra and Soesanti (2016) studied the implementation of green energy (solar home system) for the production of batik in Yogyakarta Batik SMEs. They revealed that solar home system was distributing electricity to the batik stove. Using a solar system is more economical and practical compared to using a kerosene stove. Another study suggested the framework of a sustainable batik production process (Indarti et al., 2019). Indarti et al. used literature review method from 82 articles to make the conceptual framework of sustainable batik production. There were four categories in the sustainable production of batik: designing, waxing, dying, and eliminating wax. How to treat the waste, using wax, dying process, and how to make a design of batik are the priorities on the sustainability of batik production in Indonesia. The sectors contributing to sustainable batik production are green marketing and brand quality (Wibowo et al., 2024). Product marketing strategy used environmental aspects as strategy in green marketing. Production, pricing, advertising, and sales are among the many processes that make up green marketing (Majerova, 2015). The authors found that consumer behavior in making purchase decisions was influenced by Lasem Batik's green marketing and brand quality. The authors found that it greatly affects the decision of when a buyer will buy a product, conveys good information to other customers, and buys another batik product. It makes batik products; especially the business of lasem batik will persist if it is sustained. The sustainability of batik production in Indonesia is also related to the participation of women in Indonesia. In general, batik craftsmen in the industry are women because batik craft requires special skills, especially when making written batik (Praitiwi & Darsih, 2019).

9 Sustainable Retailing and Consumption of Batik and Textile

Given the growing importance of environmental consciousness and social responsibility among consumers, sustainable batik and textile industries must adopt sustainable retailing and consumption practices. This adoption is necessary to align with consumers' evolving requirements and preferences (Jacobs et al., 2018). We examine the literature generated from sustainable consumption and sustainable retailing in the following.

9.1 Sustainable Consumption of Batik and Textile

The significance of consumption in environmental issues is widely recognized. It is widely acknowledged that consumption activities, along with other production activities, have a significant impact on various environmental issues. Understanding the profound influence of consumption on these matters, it is vital to acknowledge consumers' crucial role in shaping consumption patterns. Consumers play a significant role in environmental degradation as the main agents of consumption, as highlighted by Berglund and Matti (2006). Therefore, the role of consumers is vital in promoting sustainable consumption.

Sustainable consumption refers to a kind of consumption that takes into account the social, economic, and environmental requirements of both current and future generations (Prothero et al., 2011). According to Biswas (2017), sustainable consumption is defined as the practice of reducing resource consumption, adopting eco-friendly products, and adjusting lifestyles. Numerous variables have been identified as potential factors affecting sustainable consumption. Individual-related factors and situational/contextual factors can be distinguished among these variables. Attitudes, values, beliefs, demographic characteristics, and additional variables potentially impacting an individual's conduct are categorized as individual-related factors. In contrast, contextual-related or situational factors encompass extraneous influences that can impact the conduct of individuals such as product value, brand popularity, environment, and regulation (Kostadinova, 2016).

Žurga and Tavčer (2014) examine sustainable garment consumption from the standpoint of consumers' sustainable consumption. They highlight that just a minority of customers prioritize eco-labeling and environmental awareness. In addition, it was discovered that buyers are only prepared to pay a minor surcharge of "an additional 10%" for clothing items that have eco-labels. The authors highlight the need to increase public awareness and foster trust in eco-labeling processes to promote sustainable clothing consumption. A further study by Ciasullo et al. (2017) found that sustainable consumption plays a minor role in customers' purchasing choices. In addition, the extent to which consumers are willing to pay for environmentally friendly clothing is only marginally greater, reaching a maximum of 20%, compared to conventional clothing that is not environmentally friendly.

According to study conducted by Charista and Paramita (2023), there is a strong correlation between consumer awareness and their intention to purchase environmentally sustainable clothing in Indonesia. Indonesians are increasingly aware of the importance of sustainable clothing. Furthermore, studies have shown that there are gender differences when it comes to awareness, suggesting that women tend to be more mindful of sustainable fashion compared to men. However, there are no significant variations in average value levels. On the other hand, there is a noticeable disparity in how men and women perceive purchase intention. The results of this study have consequences that can enhance future investigations on literacy. There is a lack of comprehension and consciousness regarding sustainability in the fashion industry, particularly in comparison to other sectors. A recent study by Nuh et al. (2023) revealed exciting insights into the variations in sustainable fashion consumption among different generations in Indonesia. Younger generations, including Gen Z, are increasingly inclined to prioritize sustainability in the fashion industry. Gen Z is drawn to sustainable fashion for many reasons, such as staying on top of trends, reaping economic advantages, and prioritizing sustainability. In addition, fashion is no longer solely focused on following trends or displaying social status for Gen Z. Instead, it is about enhancing self-assurance and forming personal identities. In general, the variations in sustainable fashion consumption among different generations can be attributed to factors like age, values, and access to information (Gazzola et al., 2020).

In sustainable batik clothing consumption, according to a study by Pratama et al. (2023), six factors affect consumers' willingness to pay for batik products dyed with natural dyes. These factors include value orientation, cognitive knowledge, aesthetics, demographics, attitude toward sustainable products, and intention to purchase sustainable products. The survey also found that consumers' intention to purchase sustainable products is strongly influenced by their value system. Values can typically be categorized into two main groups: self-transcendence and self-enhancement. Self-transcendence encompasses both altruism and a drive to contribute to the betterment of society. The journey entails the exploration of one's self to achieve enhanced excellence and a broader outlook. Self-enhancement refers to individuals being driven to advance their self-interest and shape their self-perception. These values have a significant impact and catalyze action. In addition, the level of consumer knowledge also affects their willingness to purchase batik products dyed with natural dyes. This study also recognizes the significance of understanding in influencing decision-making at every stage. Knowledge is a critical factor in influencing how consumers acquire information. Consumers' intention to purchase products, especially environmentally friendly textiles, can be influenced by aesthetics, beauty, value, and knowledge. There are 15 indicators spread across seven different attributes of batik cloth. The indicators include stitching and craft quality, color durability, cloth type, ease of maintenance, dye type, batik motifs, color contrast, seller identity, eco-friendly packaging, product information, guarantees, price discounts, price-quality ratio, and customer service.

9.2 Sustainable Retailing of Batik and Textile

The retail industry has the potential to contribute significantly to promoting sustainable consumption, considering its advantageous position from the perspective of demand and supply (Paul & Rosenbaum, 2020). Retail companies can shape consumer behavior and encourage sustainable consumption patterns. Retailers can implement various initiatives to encourage sustainable consumption, such as recycling (Musa et al., 2013) and using sustainable stuff (Longoria et al., 2021). Retailers can motivate suppliers to develop environmentally friendly products (Delai & Takahashi, 2013).

With the growing emphasis on sustainable consumption, individuals are increasingly prioritizing the sustainability of fashion products and demonstrating a greater willingness to invest in sustainable options. In response to these trends, fashion retailers are making strides to establish an environmentally sustainable and socially responsible brand image. Eco-fashion is all about considering the environmental impact during the textile and clothing production process (D'Souza, 2015). Fashion products which adhere to fair trade principles, such as sweatshop-free work conditions and reduced environmental harm, are considered ethical fashion (Manchiraju & Sadachar, 2014). Emphasizing the use of sustainable or renewable materials is a key strategy in eco-fashion. Products made from sustainable materials and produced under fair-trade conditions are classified as ethical fashion. Therefore, the fashion industry needs to incorporate both eco-fashion and ethical fashion initiatives into its green branding.

According to a study by Shen and Li (2015), the product return policy in fashion retailing has a significant impact on the sustainability of retailers due to the costs associated with returns. In a study conducted by Visser et al. (2015), the researchers explore the impact of advertising strategies on consumer buying behavior. The findings suggest that advertisements that take into account the benefits to consumers and sustainable factors have the highest likelihood of encouraging purchases. In a study conducted by Manzoor et al. (2016), the researchers explore the impact of environmental and social concerns on consumers' decision-making when it comes to resource consumption. The findings suggest that by considering the diversity of consumers and their social relationships, it is possible to determine the social contexts that promote sustainable resource consumption.

Existing research also offers valuable insights for fashion companies to enhance their green marketing strategies. Lee et al. (2012) found a positive association between consumers' perceptions of green campaigns and their sustainability behavior. Consumers' green consciousness mediated this relationship. In their study, Bly et al. (2015) investigate the impact of various motivational and contextual factors on consumers' behavior toward sustainable fashion, specifically focusing on the purchase of secondhand fashion goods.

This research indicated that individual fashion preferences significantly influence customers' decisions about sustainable fashion. A study conducted by Ritch (2015) investigates the integration of sustainability practices in the fashion sector. It reveals that UK fashion retailers have embraced sustainability practices, which in turn influence consumers' decision-making. Ritch (2015) also examines how consumers perceive sustainable fashion consumption and highlights specific challenges to sustainable consumption in the fashion sector. Another study by Kumar (2014) provides an overview of the many sustainable retailing strategies in India and classifies them into nine separate categories. These encompass the encouragement of sustainable corporate practices, the execution of awareness campaigns, and the adoption of techniques to engage consumers.

In Indonesia, Batik retail has a reactive tendency and prioritizes meeting market demands rather than being proactive towards the environment, as stated by Kocak et al. (2017). Due to limited resources, batik retailers were compelled to depend solely on the current technological or marketing paths. They will only modify their behavior when there are alterations in their surroundings. Due to a scarcity of resources, they often exhibit exploitative behaviors. Batik retailers should strive to enhance their ability to identify untapped business prospects by acquiring new resources and cultivating a more inventive competitive edge (Mashahadi et al., 2016). In order to enhance their company performance, batik retail in Indonesia should focus on developing both market orientation and environmental orientation simultaneously.

10 How Sustainable Batik and Textile in Indonesia Contributes to SDGs

The Center for Crafts and Batik, Indonesian Ministry of Industry's data shows that there are 3159 batik businesses registered in Indonesia (databoks.katadata.co.id, 2024). There are 208 large and medium-scale industrial units in the batik business units, along with 2951 micro-small and medium-scale units. This micro batik business is the most numerous business unit in Indonesia compared to business units in another sector. The multifunctional role of batik in Indonesian people's lives is inseparable from the many batik industries. Apart from being everyday clothing, batik is also used as formal clothing for work, wedding ceremonies, art performances, pregnancy celebrations, carrying babies, and even at funerals. Apart from showing the variety of cultures throughout Indonesia, batik patterns and motifs themselves also reflect connections and influences from abroad, such as batik cloth with Arabic calligraphy patterns, Japanese cherry blossoms, Indian and Persian peacocks, or European flower arrangements (databoks.katadata.co.id, 2024).

Most batik SME workers in Indonesia are women. Previous research shows that there are policies made by SME owners of the batik industry for workers who are predominantly female (Kurnia, 2022). For example, there is a system of dividing work according to the abilities of workers so that workers' performance is more efficient. The dual role of women as batik industry SME workers and housewives can be overcome with a flexible working hours policy by providing a piecework system that can be taken home and done at home. The performance of these workers can be calculated from the number of batik cloths produced. In general, providing work for women in Indonesian batik MSMEs supports the UN SDGs in terms of achieving decent work and economic growth, eliminating hunger, and promoting gender equality.

In the batik dyeing process, several SMEs in Indonesia also use natural dyes. Previous studies on Batik SMEs in Semarang, Central Java, Indonesia, showed that these SMEs already used natural dyes such as mango leaves (*Mangifera indica*), suji leaves (*Dracaena angustifolia*), teak leaves (*Tectona grandis*) and tegeran (*Cudrania javanensis*) (Kamil et al., 2016). Workers use these materials to obtain the green color. Meanwhile, to produce the yellow color, batik SMEs use natural dyes from secang (*Biancaea sappan*), jackfruit peel (*Artocarpus heterophyllus*), turmeric (*Curcuma longa*) and jalawe plant (*Terminalia bellirica*). Another batik SME in the Klaten area, Central Java, Indonesia, also uses natural dyes in making written batik, namely indigo, soga (*Peltophorum pterocarpum*), tingi (*Ceriops tagal*), tegeran (*Cudrania javanensis*), and jalawe (*Terminalia bellirica*) (Chafidz & Lestari, 2021). The fixative substances used are lime and alum. The process of producing natural dyes in this SME is still performed manually, by extraction using hot water in a pan over a wood-burning stove. The extraction was repeated three times. The use of natural dyes is in line with the UN SDGs regarding good health and clean water and sanitation.

Batik SMEs in Gunungkidul, Yogyakarta, Indonesia, have also tried using the *Ganoderma lucidum* fungus to purify waste produced from the process of making batik cloth (Pratiwi et al., 2017). *G. lucidum* is a type of white-rot fungus and can grow in agro-industrial biomass. This fungus can grow in lignin with alkaline properties and has quite high laccase activity (Sitarz et al., 2013). The use of local natural sources, i.e., *G. lucidum* fungus to purify batik waste, is in line with the UN SDGs in the field of good health and clean water.

Several efforts have been made to improve the quality of batik produced by SMEs batik in Indonesia. One of the batik SMEs in Jember, East Java, Indonesia (Kurnia, 2022) regularly conducts batik cloth-making training. Meanwhile, batik SMEs in Klaten, Central Java, Indonesia, in collaboration with universities, are also conducting training on the extraction of natural dyes (Chafidz & Lestari, 2021). Extraction, which usually uses boiling water, can be varied by using other solvents (for example, ethanol) and also using a boiling-assisted extraction unit to replace a pan. Apart from that, the size of natural raw materials can also be reduced by using a crusher and sieving machine because the particle size affects the extraction results of natural dyes. These training activities for SME workers are in line with UN SDGs in the field of education.

11 Managerial Recommendations of Batik and Textile in Indonesia

From the description of the batik and textile industry in Indonesia above, we found that several batik SMEs in Indonesia already support UN SDGs programs such as gender equality, clean water, education, and good health. However, not all batik industries in Indonesia have implemented this, so further steps are needed so that all batik industry processes can be in line with the UN SDGs program. In order for the batik industry in Indonesia to be sustainable and balanced in terms of social, economic, and environmental aspects, it is necessary to make substantial efforts. Of course, to achieve this goal requires participation from all stakeholders, including the government, batik industry business players, sellers, consumers, and investors. The Indonesian government can support the batik industry to be in line with the UN SDGs, for example, with policies in the field of taxation, so that the batik industry, especially home industries and SMEs, is not too burdened by taxes. The welfare of the batik industry in Indonesia, particularly SMEs, will be enhanced, and this is in line with the UN SDGs goal of eliminating poverty and hunger. The batik industry, whether large or small, can utilize natural materials to reduce the waste. Apart from that, it is hoped that we can use technological innovation so that we can carry out more efficient and sustainable batik industrial processes, as well as implement an adequate wastewater treatment system to handle the waste produced before it is discharged into rivers and waters. The batik industry must also be able to create a sustainable concept for the batik products produced so that they can be recycled by consumers. Retailers are expected to be able to create sustainable business models such as eco-labeling for batik products, possible rental service facilities for batik products, and reducing carbon emissions in the sales process. Consumers are expected to have awareness and ability in sustainable batik consumption, including using batik products as long as possible and recycling them if it can still be done. Batik industry investors can carry out a comprehensive evaluation of the batik industry supply chain, have a forward view regarding long-term investments such as new technology, and have the awareness to provide new opportunities in the field of sustainable batik products.

12 Conclusion and Future Research

The aim of this study is to examine how the batik and textile industries can achieve the SDGs goals and maintain the sustainable production of the industries. We categorized the 17 goals of SDGs from the UN as economic, social, and environmental. This study proposed that there are priority sectors contributed to achieve SDGs, i.e., green industry standard, green marketing, and quality of product. The balance of the three factors from economic, social, and environmental perspectives is currently not balanced. Currently, the batik and textile industry still focus on the economic and environmental sectors. Recommendations actions for batik and textile production in Indonesia create guidelines for the policymakers and stakeholders.

For future research, we suggest that research should focus on balancing three sectors: economic, social, and environmental, in batik and textile production. Collaboration between stakeholders such as policymakers, industries, consumers, and other stakeholders to achieve SDGs are more explored. More attention should be paid to SDS goals such as "No poverty," "Peace, Justice and Strong Institutions," "Quality Education, Life below Water," and "Life in the Land." Future research should be focused on some issues, such as how to decrease the poverty rate in an area where the batik and textile industry is established, how to make more stable and sustainable the batik and textile industries, how to manage the waste and protect marine life from the chemicals, plastic, and solid waste that industries produce, and how to protect biodiversity against climate change. Use of several methods in analyzing the sustainability of batik and textile production.

References

Abdulqader, E., Abdullah, S. R. S., Hasan, H. A., Ismail, N. I., Rahim, R. A. A., & Idris, M. (2019). Plant-assisted remediation of wastewater contaminated with methyl orange using *Scirpus grossus*. *Journal of Environmental Biology, 40*, 515–523. https://doi.org/10.22438/eb/40/3(SI)/Sp-16

Ahi, P., & Searcy, C. (2015). Measuring social issues in sustainable supply chains. *Measuring Business Excellence, 19*(1), 33–45. https://doi.org/10.1108/MBE-11-2014-0041

Ahmad, A. L., Harris, W. A., Syafiee, & Seng, O. B. (2002). Removal of dye from wastewater of textile industry using membrane technology. *Jurnal Teknologi, 36*(F), 31–44. https://doi.org/10.11113/jt.v36.581

Ajizah, S., Hidayat, N., & Suhartini, S. (2021). Modelling green production process in the natural dyes batik industry using cleaner production options. *Research Journal of Industrial Pollution Prevention Technology, 12*(1), 39–54. https://doi.org/10.21771/jrtppi.2021.v12.no1

Ali, S., Hussain, T., & Nawaz, R. (2009). Optimization of alkaline extraction of natural dye from Henna leaves and its dyeing on cotton by exhaust method. *Journal of Cleaner Production, 17*(1), 61–66. https://doi.org/10.1016/j.jclepro.2008.03.002

Ali, S., Jabeen, S., Hussain, T., Noor, S., & Siddiqua, U. H. (2016). Optimization of extraction condition of natural dye from pomegranate peels using response surface methodology. *International Journal of Engineering Science Technologies, 5*(7), 542–548. https://doi.org/10.5281/zenodo.57042

Ariyanti, M., & Asbur, Y. (2018). The tarum plant (*Indigofera tinctoria* Linn.) produces dyes. *Small Islands Forest Journal, 2*(1), 109–122. Retrieved from https://ojs3.unpatti.ac.id/index.php/jhppk/article/view/754

Arumsari, A. (2020). *Application of design ethics in the middle class fashion industry in Bali* (pp. 20–23). Bandung Institute of Technology.

Atika, V., & Haerudin, A. (2013). Effect of natural resin composition on wax melting temperature for natural color batik. *Dynamics of Crafts and Batik Journal, 30*(1), 23–29.

Atika, V., & Salma, I. R. (2017). Staining quality of tegeran wood (*Cudrania javanensis*) in batik. *Dynamics of Crafts and Batik Journal, 34*(1), 11–18.

Atilgan, T. (2007). Eco-labelling applications in the textile & apparel sector in Turkey. *Fibres and Textiles in Eastern Europe, 15*(2), 14–19.

Bechtold, T., Turcanu, A., Ganglberger, E., & Geissler, S. (2003). Natural dyes in modern textile dyehouses—How to combine experiences of two centuries to meet the demands of the future? *Journal of Cleaner Production, 11*(5), 499–509. https://doi.org/10.1016/S0959-6526(02)00077-X

Berglund, C., & Matti, S. (2006). Citizen and consumer: The dual role of individuals in environmental policy. *Environmental Politics, 15*(4), 550–571. https://doi.org/10.1080/09644010600785176

Biswas, A. (2017). A consumption value-gap analysis for sustainable consumption. *Environmental Science and Pollution Research, 24*(8), 7714–7725. https://doi.org/10.1007/s11356-016-8355-9

Bly, S., Gwozdz, W., & Reisch, L. A. (2015). Exit from the high street: An exploratory study of sustainable fashion consumption pioneers. *International Journal of Consumer Studies, 39*(2), 125–135. https://doi.org/10.1111/ijcs.12159

Bottani, E., Tebaldi, L., Lazzari, I., & Casella, G. (2020). Economic and environmental sustainability dimensions of a fashion supply chain: A quantitative model. *Production, 30*(3). https://doi.org/10.1590/0103-6513.20190156

Brammer, S., & Walker, H. (2011). Sustainable procurement in the public sector: An international comparative study. *International Journal of Operations and Production Management, 31*(4), 452–476. https://doi.org/10.1108/01443571111119551

Cai, Y. J., & Choi, T. M. (2020). A United Nations' Sustainable Development Goals perspective for sustainable textile and apparel supply chain management. *Transportation Research Part E: Logistics and Transportation Review, 141*(July), 102010. https://doi.org/10.1016/j.tre.2020.102010

Chafidz, A., & Lestari, A. Y. D. (2021). Introduction to natural dye extraction technology for dyes. *Community Journal: Journal of Community Service, 3*(2), 101–108. https://ojs.stiami.ac.id/index.php/jks/article/view/1271

Charista, V., & Paramita, E. L. (2023). Indonesian customer awareness toward purchase intention of sustainable fashion. *Proceedings of the 2nd International Interdisciplinary Conference on Environmental Sciences and Sustainable Developments Education and Green Economy,*

Advances in Social Science, Education and Humanities Research, 794, 402–407. https://doi.org/10.2991/978-2-38476-172-2_67
Chi, T. (2011). Building a sustainable supply chain: An analysis of corporate social responsibility (CSR) practices in the Chinese textile and apparel industry. *Journal of the Textile Institute, 102*(10), 837–848. https://doi.org/10.1080/00405000.2010.524361
Choi, T. M., & Cai, Y. J. (2018). Impacts of lead time reduction on fabric sourcing in apparel production with yield and environmental considerations. *Annals of Operations Research, 290*, 521–542. https://doi.org/10.1007/s10479-018-2838-7
Choi, T. M., Cai, Y. J., & Shen, B. (2019). Sustainable fashion supply chain management: A system of systems analysis. *IEEE Transactions on Engineering Management, 66*(4), 730–745. https://doi.org/10.1109/TEM.2018.2857831
Ciasullo, M. V., Maione, G., Torre, C., & Troisi, O. (2017). What about sustainability? An empirical analysis of consumers' purchasing behavior in fashion context. *Sustainability (Switzerland), 9*(9), 1–18. https://doi.org/10.3390/su9091617
Collins, E., Roper, J., & Lawrence, S. (2010). Sustainability practices: Trends in New Zealand businesses. *Business Strategy and the Environment, 19*, 479–494. https://doi.org/10.1002/bse.653
D'Souza, C. (2015). Marketing challenges for an eco-fashion brand: A case study. *Fashion Theory – Journal of Dress Body and Culture, 19*(1), 67–82. https://doi.org/10.2752/175174115X14113933306824
Darsih, C., Lestari, D. W., Pratiwi, D., Indrianingsih, A. W., Rosyida, V. T., Apriyana, W., Nisa, K., Hayati, S. N., Handayani, S., & Wulanjati, M. P. (2019). Dyeing of cotton fabric with natural dye from *Cudrania javanensis* using soka (*Ixora javanica*) leaves extract as bio-mordant. *Dynamics of Crafts and Batik Journal, 36*(2), 105–112. https://doi.org/10.22322/dkb.V36i1.4149
Daud, N. M., Abdullah, S. R. S., Hasan, H. A., Ismail, N. I., & Dhokhikah, Y. (2022). Integrated physical-biological treatment system for batik industry wastewater: A review on process selection. *Science of the Total Environment, 819*, 152931. https://doi.org/10.1016/j.scitotenv.2022.152931
Daud, N. M., Abdullah, S. R. S., Hasan, H. A., Othman, A. R., & Ismail, N. I. (2023). Coagulation-flocculation treatment for batik effluent as a baseline study for the upcoming application of green coagulants/flocculants towards sustainable batik industry. *Heliyon, 9*(6), e17284. https://doi.org/10.1016/j.heliyon.2023.e17284
Delai, I., & Takahashi, S. (2013). Corporate sustainability in emerging markets: Insights from the practices reported by the Brazilian retailers. *Journal of Cleaner Production, 47*, 211–221. https://doi.org/10.1016/j.jclepro.2012.12.029
Denyer, D., & Tranfield, D. (2009). Producing a systematic review. In *The Sage handbook of organizational research methods* (pp. 671–689). Sage Publications Ltd.
Diabat, A., Kannan, D., & Mathiyazhagan, K. (2014). Analysis of enablers for implementation of sustainable supply chain management – A textile case. *Journal of Cleaner Production, 83*, 391–403. https://doi.org/10.1016/j.jclepro.2014.06.081
Dissanayake, G., & Sinha, P. (2015). An examination of the product development process for fashion remanufacturing. *Resources, Conservation and Recycling, 104*, 94–102. https://doi.org/10.1016/j.resconrec.2015.09.008
Ellram, L. M., & Cooper, M. C. (2014). Supply chain management: It's all about the journey, not the destination. *Journal of Supply Chain Management, 50*(1), 8–20. https://doi.org/10.1111/jscm.12043
Fadhilah, S. N., & Widiawati, D. (2021). Development of the Ciwaringin Batik production process using sustainable design analysis. *Proceedings of the National Seminar on the Handicraft and Batik Industry, 3*(1), A.09, 1–18. Retrieved from https://proceeding.batik.go.id/index.php/SNBK/article/view/99

Garg, A., Bhat, K. L., & Bock, C. W. (2002). Mutagenicity of aminoazobenzene dyes and related structures: A QSAR/QPAR investigation. *Dyes and Pigments, 55*(1), 35–52. https://doi.org/10.1016/S0143-7208(02)00070-0

Gazzola, P., Pavione, E., Pezzetti, R., & Grechi, D. (2020). Trends in the fashion industry. The perception of sustainability and circular economy: A gender/generation quantitative approach. *Sustainability (Switzerland), 12*(7), 1–19. https://doi.org/10.3390/su12072809

Gómez-Luciano, C. A., Rondón Domínguez, F. R., González-Andrés, F., De Meneses, U. L., & B. (2018). Sustainable supply chain management: Contributions of supplies markets. *Journal of Cleaner Production, 184*, 311–320. https://doi.org/10.1016/j.jclepro.2018.02.233

Gopal, P. R. C., & Thakkar, J. (2016). Sustainable supply chain practices: An empirical investigation on Indian automobile industry. *Production Planning and Control, 27*(1), 49–64. https://doi.org/10.1080/09537287.2015.1060368

Gray, C., & Charter, M. (2007). Remanufacturing and product design. *International Journal of Product Development, 6*(3–4), 375–392. https://cfsd.org.uk/Remanufacturing and Product Design.pdf

Gunawan, A. A., Bloemer, J., van Riel, A. C. R., & Essers, C. (2022). Institutional barriers and facilitators of sustainability for Indonesian Batik SMEs: A policy agenda. *Sustainability (Switzerland), 14*(14), 1–25. https://doi.org/10.3390/su14148772

Gupta, S., & Palsule-Desai, O. D. (2011). Sustainable supply chain management: Review and research opportunities. *IIMB Management Review, 23*(4), 234–245. https://doi.org/10.1016/j.iimb.2011.09.002

Haji, A. (2010). Functional dyeing of wool with natural dye extracted from *Berberis vulgaris* wood and *Rumex hymenosepolus* root as biomordant. *Iranian Journal Chemistry & Chemical Engineering, 29*(3), 55–60. https://doi.org/10.30492/ijcce.2010.6513

Handayani, F. (2009). *Getting to Know and Making Batik (Indonesian language). Edition 2.* Buana Cipta Pustaka.

Handayani, W., Kristijanto, A. I., & Hunga, A. I. R. (2018). Are natural dyes eco-friendly? A case study on water usage and wastewater characteristics of batik production by natural dyes application. *Sustainable Water Resources Management, 4*(4), 1011–1021. https://doi.org/10.1007/s40899-018-0217-9

Haque, M. A., Khan, G. M. A., Razzaque, S. M. A., Khatun, K., Chakraborty, A. K., & Alam, M. S. (2013). Extraction of rubiadin dye from swietenia mahagoni and its dyeing characteristics onto silk fabric using metallic mordants. *Indian Journal of Fibre & Textile Research, 38*, 280–284. https://nopr.niscpr.res.in/bitstream/123456789/21433/1/IJFTR%2038(3)%20280-284.pdf

Hasibuan, S., & Hidayati, J. (2018). The Integration of cleaner production innovation and creativity for supply chain sustainability of bogor batik SMEs. *International Journal of Industrial and Manufacturing Engineering, 12*(5), 679–684. https://doi.org/10.5281/zenodo.1474275

Hsueh, C. F. (2015). A bilevel programming model for corporate social responsibility collaboration in sustainable supply chain management. *Transportation Research Part E: Logistics and Transportation Review, 73*, 84–95. https://doi.org/10.1016/j.tre.2014.11.006

https://databoks.katadata.co.id/datapublish/2022/10/02/usaha-batik-indonesia-mayoritas-berskala-mikro, accessed on April 5, 2024.

https://www.fwd.co.id/id/fwdmax/passionstory-fashion-and-music/beda-antara-batik-tulis-cap-dan-printing-sudah-tahu/, accessed on April 6, 2024.

https://www.inditex.com/itxcomweb/en/home, accessed on April 4, 2024.

https://www.news.palmoilmagazine.com/news/9699/indonesia%2D%2D-s-batik-industry-commits-to-using-sustainable-palm-oil-https://www.palmoilmagazine.com/rspo/2023/11/25/laweyan-sustainable-palm-batik-started-to-be-marketed-in-rspo-rt2023/, accessed on April 4, 2024.

Indarti, Rahayu I. A. T., & Peng L. H. (2019). Sustainable batik production: Review and research framework. Advance in Social science, education and humanities research, 390

https://www.un.org/sustainabledevelopment/sustainable-development-goals/, accessed on April 4, 2024.

Indrayani, L. (2019). Strategic efforts to manage batik industry waste in realizing environmentally friendly batik. *Proceedings of the National Seminar on the Handicraft and Batik Industry, 1*(1). Retrieved from https://proceeding.batik.go.id/index.php/SNBK/article/view/34

Indrianingsih, A. W., & Darsih, C. (2013). Natural dyes from plants extract and its applications in indonesian textile small medium scale enterprise. *Eksergi, 11*(1), 16–22. https://doi.org/10.31315/e.v11i1.327

Indrianingsih, A. W., Rosyida, V. T., Darsih, C., & Apriyana, W. (2021). Antibacterial activity of *Garcinia mangostana* peel-dyed cotton fabrics using synthetic and natural mordants. *Sustainable Chemistry and Pharmacy, 21*, 100440. https://doi.org/10.1016/j.scp.2021.100440

Industrial Law (Indonesian Language), Indonesia, UU Nomor 3 Tahun. (2014). https://peraturan.bpk.go.id/Details/38572/uu-no-3-tahun-2014, accessed on April 5, 2024.

Jacobs, K., Petersen, L., Hörisch, J., & Battenfeld, D. (2018). Green thinking but thoughtless buying? An empirical extension of the value-attitude-behaviour hierarchy in sustainable clothing. *Journal of Cleaner Production, 203*, 1155–1169. https://doi.org/10.1016/j.jclepro.2018.07.320

Kalkanci, B., & Plambeck, E. L. (2020). Managing supplier social and environmental impacts with voluntary versus mandatory disclosure to investors. *Management Science, INFORMS, 66*(8), 3311–3328. https://doi.org/10.1287/mnsc.2019.3382

Kamil, A., Bakhtiar, A., & Sriyanto, S. (2016). Selection of natural dyes for hand-written batik in Semarang small and medium enterprises using the Analytical Hierarchy Process (AHP) method (Case study of Semarang Batik SMEs). *Industrial Engineering Online Journal, 5*(2), 1–7. https://ejournal3.undip.ac.id/index.php/ieoj/article/view/11920/11574

Kanchana, R., Fernandes, A., Bhat, B., Bhurkule, S., Dessai, S., & Mohan, R. (2013). Dyeing of textiles with sarwo adi dyes—An ecofriendly approach. *International Journal of ChemTech Research, 5*(5), 2102–2109. https://sphinxsai.com/2013/JulySept13/chPDF/CT=06(2102-2109)JS13.pdf.

Kant, R. (2012). Textile dyeing industry an environmental hazard. *Natural Science, 4*(1), 22. https://doi.org/10.4236/ns.2012.41004

Kocak, A., Carsrud, A., & Oflazoglu, S. (2017). Market, entrepreneurial, and technology orientations: Impact on innovation and firm performance. *Management Decision, 55*(2), 248–270. https://doi.org/10.1108/MD-04-2015-0146

Koksal, D., Strahle, J., Muller, M., & Freise, M. (2017). Social sustainable supply chain management in the textile and apparel industry—A literature review. *Sustainability, 9*(1), 100. https://doi.org/10.3390/su9010100

Kostadinova, E. (2016). Sustainable consumer behaviour: A literature review. *Economic Alternatives, 2*(2), 224–234. https://ideas.repec.org/a/nwe/eajour/y2016i2p224-234.html

Kumar, P. (2014). Greening retail: An indian experience. *International Journal of Retail and Distribution Management, 42*(7), 613–625. https://doi.org/10.1108/IJRDM-02-2013-0042

Kurnia, A. M. (2022). Productivity of female workers in increasing the number of batik production in umkm kh batik andongsari jember. *Undergraduate thesis,* Universitas Islam Negeri Kiai Haji Achmad Siddiq, Jember. http://digilib.uinkhas.ac.id/12267/

Kurniawan, C., Mahatmanti, F. W., Mahfudz, J., Qonita, U., Nugraha, R., & Bahron, H. (2021). Color and chemical oxygen demand removal of *Strobilanthes cusia* fermentation and natural indigo-based batik dyeing wastewater. *International Journal of Research Innovation and Entrepreneurship, 2*(1), 10–16. https://doi.org/10.15294/ijrie.v2i1.48688

Kusumastuti, A., Atika, Achmadi, T. A., Phusavat, K., & Hidayanto, A. N. (2023). Assessment of producer's perspective on the production of environmentally friendly fashion products: A case study in Indonesian natural dyes batik craftsmen. *Environmental Science and Pollution Research, 30*, 124767–124779. https://doi.org/10.1007/s11356-022-23330-z

Kusumawati, N., Kistyanto, A., & Samik. (2019). Exploration of natural dyes by using a combination of Caesalpinia sappan and Leucaena leucocephala L. leaves. *Proceedings of the National Seminar on Chemistry, Atlantis Highlights in Chemistry and Pharmaceutical Sciences, 1*, 33–37. https://doi.org/10.2991/snk-19.2019.9

Kusumawati, N., Sianita, M. M., Santosa, A. B., Muslim, S., & Auliya, A. R. (2021). Development of eco-friendly dyeing process based on caesalpinia sappan L. bark, Cocos nucifera fiber and Leucaena leucocephala leaves. *International Journal on Advanced Science, Engineering and Information Technology, 11*(5), 1964–1976. https://doi.org/10.18517/ijaseit.11.5.13400

Lee, N., Choi, Y. J., Youn, C., & Lee, Y. (2012). Does green fashion retailing make consumers more eco-friendly?: The influence of green fashion products and campaigns on green consciousness and behavior. *Clothing and Textiles Research Journal, 30*(1), 67–82. https://doi.org/10.1177/0887302X12446065

Lestari, D., Atika, V., Isnaini, I., Haerudin, A., & Arta, T. (2020a). Effect of extraction ph on silk batik coloring using natural mahogany bark dye (*Switenia mahagoni*). *Journal of Process Engineering, 14*(1), 74–81. https://doi.org/10.22146/jrekpros.54439

Lestari, D., Atika, V., Satria, Y., Fitriani, A., & Susanto, T. (2020b). Application of tannin mordant in dyeing cotton batik fabric using high natural colors (Ceriops tagal). *Journal of Process Engineering, 14*(2), 128–136. https://doi.org/10.22146/jrekpros.57891

Longoria, L. C., López-Forniés, I., Sáenz, D. C., & Sierra-Pérez, J. (2021). Promoting sustainable consumption in higher education institutions through integrative co-creative processes involving relevant stakeholders. *Sustainable Production and Consumption, 28*, 445–458. https://doi.org/10.1016/j.spc.2021.06.009

Lun, Y. E., Abdullah, S. R. S., Hasan, H. A., Othman, A. R., Kurniawan, S. B., Imron, M. F., Al Falahi, O. A., Said, N. S. M., Sharuddin, S. S. N., & Ismail, N. I. (2022). Integrated emergent-floating planted reactor for textile effluent: Removal potential, optimization of operational conditions and forthcoming waste management strategy. *Journal of Environmental Management, 311*, 114832. https://doi.org/10.1016/j.jenvman.2022.114832

Luthra, S., Garg, D., & Haleem, A. (2015). Critical success factors of green supply chain management for achieving sustainability in Indian automobile industry. *Production Planning and Control, 26*(5), 339–362. https://doi.org/10.1080/09537287.2014.904532

Majerova, J. (2015). Analysis of Solvak consumer's perception of the green marketing activities. *Procedia Economics and Finance, 26*(15), 553–560. https://doi.org/10.1016/s2212-5671(15)00954-5

Manchiraju, S., & Sadachar, A. (2014). Personal values and ethical fashion consumption. *Journal of Fashion Marketing and Management, 18*(3), 357–374. https://doi.org/10.1108/JFMM-02-2013-0013

Maniu, I., Costache, C., & Dumitraşcu, D. D. (2021). Adoption of green environmental practices in small and medium-sized enterprises: Entrepreneur and business policies patterns in Romania. *Sustainability (Switzerland), 13*(9). https://doi.org/10.3390/su13094968

Manzoor, T., Rovenskaya, E., & Muhammad, A. (2016). Game-theoretic insights into the role of environmentalism and social-ecological relevance: A cognitive model of resource consumption. *Ecological Modelling, 340*, 74–85. https://doi.org/10.1016/j.ecolmodel.2016.09.007

Mashahadi, F., Ahmad, N. H., & Mohamad, O. (2016). Market orientation and innovation ambidexterity: A synthesized model for internationally operated herbal-based small and medium enterprises (HbSMEs). *Procedia Economics and Finance, 37*(16), 145–151. https://doi.org/10.1016/s2212-5671(16)30105-8

Musa, H. M., Hayes, C., Bradley, M. J., Clayson, A., & Gillibrand, G. (2013). Measures aimed at reducing plastic carrier bag use: A consumer behaviour focused study. *Natural Environment, 1*(1), 17. https://doi.org/10.12966/ne.06.02.2013

Na, Y., & Na, D. K. (2015). Investigating the sustainability of the Korean textile and fashion industry. *International Journal of Clothing Science and Technology, 27*(1), 22–33. https://doi.org/10.1108/IJCST-08-2013-0085

Nepstad, D., Irawan, S., Bezerra, T., Boyd, W., Stickler, C., Shimada, J., Carvalho, O., Macintyre, K., Dohong, A., Alencar, A., Azevedo, A., Tepper, D., & Lowery, S. (2013). More food, more forests, fewer emissions, better livelihoods: Linking REDD+, sustainable supply chains and domestic policy in Brazil, Indonesia and Colombia. *Carbon Management, 4*(6), 639–658. https://doi.org/10.4155/cmt.13.65

Nindita, V., Purwanto, D., & Sutrisnanto, D. (2012). Evaluation of eco-efficiency implementation in one of the batik small and medium enterprises in Pekalongan regency (In Indonesian language). *Journal of Industrial Pollution Prevention Technology Research, 2*(2), 82–91.

Nisrina, N., & Ardi, R. (2022). The developing key performance indicators for sustainable supply chain in indonesian fashion industry. In *3rd African international conference on industrial engineering and operations management* (pp. 1555–1565). https://doi.org/10.46254/af03.20220375

Noerhartati, E., Soesatyo, Y., Moedjito, M., & T.C., & Khamidi, A. (2020). Determinants of sustainable supply chain performance: The role of corporate entrepreneurship in indecision textile companies. *International Journal of Supply Chain Management, 9*(1), 106–112. https://doi.org/10.59160/ijscm.v9i1.4282

Nuh, A., Munir, M. M., & Muhibban, M. (2023). Fashion engagement and pro-environmental attitudes: Drivers of sustainable fashion consumption in Indonesia. *Journal of Enterprise and Development, 5*(3), 460–478. https://doi.org/10.20414/jed.v5i3.7292

Nurdalia, I. (2006). Study and analysis of opportunities for implementing cleaner production in small batik stamped businesses: Case study of three small stamped batik industrial businesses in Pekalongan. *[Thesis]. Master's Program in Environmental Science*, Universitas Diponegoro. Semarang. https://core.ac.uk/download/pdf/11715631.pdf

Oelze, N. (2017). Sustainable supply chain management implementation-enablers and barriers in the textile industry. *Sustainability (Switzerland), 9*(8). https://doi.org/10.3390/su9081435

Özceylan, E., Demirel, N., Çetinkaya, C., & Demirel, E. (2017). A closed-loop supply chain network design for automotive industry in Turkey. *Computers and Industrial Engineering, 113*, 727–745. https://doi.org/10.1016/j.cie.2016.12.022

Parisi, M. L., Fatarella, E., Spinelli, D., Pogni, R., & Basosi, R. (2015). Environmental impact assessment of an eco-efficient production for coloured textiles. *Journal of Cleaner Production, 108*, 514–524. https://doi.org/10.1016/j.jclepro.2015.06.032

Paul, J., & Rosenbaum, M. (2020). Retailing and consumer services at a tipping point: New conceptual frameworks and theoretical models. *Journal of Retailing and Consumer Services, 54*(October 2019), 101977. https://doi.org/10.1016/j.jretconser.2019.101977

Pertiwi, C. D., Irawati, A., & Chen, J. F. (2020). The effect of corporate social responsibility disclosure on financial performance: Comparative study of textile companies listed on Indonesia and Taiwan stock exchanges. In *Proceeding of 6th international conference on information technology and business* (pp. 17–25) https://jurnal.darmajaya.ac.id/index.php/icitb/article/view/2532

Praitiwi, D., & Darsih, C. (2019). Training of natural dyes production and its application for batik at batik manggar gasing SMEs. *Journal of MIPA Community Service and Mathematics and Natural Sciences Education, 3*(2), 53–57. https://doi.org/10.21831/jpmmp.v3i2.22935

Prasad, D. S., Pradhan, R. P., Gaurav, K., & Sabat, A. K. (2020). Critical success factors of sustainable supply chain management and organizational performance: An exploratory study. *Transportation Research Procedia, 48*, 327–344. https://doi.org/10.1016/j.trpro.2020.08.027

Pratama, A. Z., Liquiddanu, E., & Pujiyanto, E. (2023). Factors affecting the consumers' willingness to pay for sustainable fashion products (study case: Natural dyed batik). In *Proceedings of the 8th North American international conference on industrial engineering and operations management* (pp. 985–992). https://doi.org/10.46254/na8.20230252

Pratiwi, D., Indrianingsih, A. W., Darsih, C., & Hernawan. (2017). Decolorization and degradation of batik dye effluent using Ganoderma lucidum. *IOP Conference Series: Earth and Environmental Science, 101*, 012034. https://doi.org/10.1088/1755-1315/101/1/012034

Prothero, A., Dobscha, S., Freund, J., Kilbourne, W. E., Luchs, M. G., Ozanne, L. K., & Thøgersen, J. (2011). Sustainable consumption: Opportunities for consumer research and public policy. *International Journal of the Economics of Business, 18*(2), 155–160. https://doi.org/10.1080/13571516.2011.584419

Pujilestari, T. (2015). Review: Sources and use of natural dyes for industrial needs. *Dynamics of Crafts and Batik, 32*(2), 93–106. https://doi.org/10.22322/dkb.v32i2.1365

Puslitbangbun. (2019). *National Seminar on Natural Fibers*. Retrieved from http://perkebunan.litbang.pertanian.go.id/seminar-nasional-serat-alam/. Accessed on April 1, 2024.

Qi, Y., Zhao, X., & Sheu, C. (2011). The impact of competitive strategy and supply chain strategy on business performance: The role of environmental uncertainty. *Decision Sciences, 42*(2), 371–389. https://doi.org/10.1111/j.1540-5915.2011.00315.x

Rausch-Phan, M. T., & Siegfried, P. (2022). Conclusion of sustainable supply chain management: Learning from the German automotive industry. In *Sustainable supply chain management. Business guides on the go*. Springer. https://doi.org/10.1007/978-3-030-92156-9_6

Regulation of the Minister of Industry on Green Industry Standards for the Batik Industry (Indonesian language), Indonesia, Permen Perindustrian Nomor 10 Tahun (2023). https://peraturan.bpk.go.id/Details/260278/permenperin-no-10-tahun-2023, accessed on April 5, 2024.

Ritch, E. L. (2015). Consumers interpreting sustainability: Moving beyond food to fashion. *International Journal of Retail & Distribution Management, 43*(12), 1162–1181. https://doi.org/10.1108/IJRDM-04-2014-0042

Riyanto, O. A. W., Siswadi, & Mardiana F. (2018). Implementation of green manufacturing in traditional batik sme to reduce waste of raw materials and waste water. *International Journal of Scinece and Engineering Investigation, 7*, 76.

Rizos, V., Behrens, A., van der Gaast, W., Hofman, E., Ioannou, A., Kafyeke, T., Flamos, A., Rinaldi, R., Papadelis, S., Hirschnitz-Garbers, M., & Topi, C. (2016). Implementation of circular economy business models by small and medium-sized enterprises (SMEs): Barriers and enablers. *Sustainability (Switzerland), 8*(11). https://doi.org/10.3390/su8111212

Rosyida, A., & Zulfiya, A. (2013). Coloring textile materials using jackfruit wood extract and coloring techniques to get optimal results. *Journal of Process Engineering, 7*(2), 52–58. https://doi.org/10.22146/jrekpros.4952

Sarasi, V., Primiana, I., Harsanto, B., & Satyakti, Y. (2023). Sustainable supply chain of Indonesia's textile & apparel industry: Opportunities and challenges. *Research Journal of Textile and Apparel, 28*, 819. https://doi.org/10.1108/RJTA-08-2022-0091

Sari, D. P., Hartini, S., Rinawati, D. I., & Wicaksono, T. S. (2012). Measuring eco-efficiency levels using life cycle assessment to create sustainable production in the small and medium batik industry. *Journal of Industrial Engineering, 14*(2), 137–144. https://doi.org/10.9744/jti.14.2.137-144

Seman, N. A. A., Zakuan, N., Jusoh, A., Arif, M. S. M., & Saman, M. Z. M. (2012). Green supply chain management: A review and research direction. *International Journal of Managing Value and Supply Chains, 3*(1), 1–18. https://doi.org/10.5121/ijmvsc.2012.3101

Seuring, S., & Müller, M. (2008). From a literature review to a conceptual framework for sustainable supply chain management. *Journal of Cleaner Production, 16*(15), 1699–1710. https://doi.org/10.1016/j.jclepro.2008.04.020

Shaharuddin, S. I. S., Shamsuddin, M. S., Drahman, M. H., Hasan, Z., Mohd Asri, N. A., Nordin, A. A., & Shaffiar, N. M. (2021). A review on the Malaysian and Indonesian batik production, challenges, and innovations in the 21st century. *SAGE Open, 11*(3). https://doi.org/10.1177/21582440211040128n

Shen, B., & Li, Q. (2015). Impacts of returning unsold products in retail outsourcing fashion supply chain: A sustainability analysis. *Sustainability (Switzerland), 7*(2), 1172–1185. https://doi.org/10.3390/su7021172

Shofari, F. D., Kurniawati, D. A., & Paramawardhani, H. (2022). Green productivity approach in batik industry. *IOP Conference Series: Earth and Environmental Science, 1039*, 12009. https://doi.org/10.1088/1755-1315/1039/1/012009

Siddique, K., Rizwan, M., Shahid, M. J., Ali, S., Ahmad, R., & Rizvi, H. (2017). Textile wastewater treatment options: A critical review. In N. Anjum, S. Gill, & N. Tuteja (Eds.), *Enhancing cleanup of environmental pollutants*. Springer. https://doi.org/10.1007/978-3-319-55423-5_6

Siswiyanti, S., Rusnoto, R., Luthfianto, S., & Nurjanah, N. (2019). The application of quality function deployment (qfd) towards the design of batik coloring machines to increase the value of

fading and stains on fabrics. *Industrial Engineering Scientific Journal, 18*(1), 95–102. https://doi.org/10.23917/jiti.v18i1.7165
Sitarz, A. K., Mikkelsen, J. D., Højrup, P., & Meyer, A. S. (2013). Identification of a laccase from *Ganoderma lucidum* CBS 229.93 having potential for enhancing cellulase catalyzed lignocellulose degradation. *Enzyme and Microbial Technology, 53*, 378–385. https://doi.org/10.1016/j.enzmictec.2013.08.003
Soykoth, M. W., Rahaman, M. M., Hossain, M. A., & Alam, M. J. (2022). Green supply chain and customer satisfaction in modern retailing: A dual-perception analysis with SEM approach. *International Journal of Integrated Supply Management, 15*(4), 381–409. https://doi.org/10.1504/ijism.2022.125990
Srisawat, S., & Srisawat, N. (2020). The effect of green supply chain management practices on the sustainable performance of the textile industry. *International Journal of Supply Chain Management, 9*(2), 300–308. https://doi.org/10.3390/su15086677
Subagyo, P. K., & Soelistyowati, S. (2023). The effect of synthetic dyes on batik fabric coloring. *Folio, 2*(2). https://journal.uc.ac.id/index.php/FOLIO/article/view/3476
Suparno, O. (2020). Potential and future of Indonesian natural fibers as raw materials for various industries. *Journal of Agricultural Industrial Technology, 30*(2), 221–227. https://doi.org/10.24961/j.tek.ind.pert.2020.30.2.221
Susanto, S. (1980). *Arts and crafts and batik* (1st ed., Indonesian language ed.). Balai Penelitian Batik dan Kerajinan.
Sutisna, E., Wibowo, M., Rokhmat, D. Y., Rahman, R., Murniati, M., & Abdullah, K. (2017). Batik wastewater treatment using TiO2 nanoparticles coated on the surface of plastic sheet. *Procedia Engineering, 170*, 78–83. https://doi.org/10.1016/j.proeng.2017.03.015
Swarnakar, V., Singh, A. R., Antony, J., Tiwari, A. K., & Cudney, E. (2021). Development of a conceptual method for sustainability assessment in manufacturing. *Computers and Industrial Engineering, 158*(November 2020), 107403. https://doi.org/10.1016/j.cie.2021.107403
Syahputra, R., & Soesanti, I. (2016). Application of green energy for batik production process. *Journal of Theoretical and Applied Information Technology, 91*(2). http://www.jatit.org/volumes/Vol91No2/4Vol91No2.pdf
Tajbakhsh, A., & Hassini, E. (2015). A data envelopment analysis approach to evaluate sustainability in supply chain networks. *Journal of Cleaner Production, 105*, 74–85. https://doi.org/10.1016/j.jclepro.2014.07.054
Tanzil, M. Y. (2017). The sustainable practice of Indonesian fashion brands. *International Conference on Art, Craft, Culture, and Design*. Bandung. pp. 203–211. https://media.neliti.com/media/publications/349116-the-sustainable-practices-of-indonesian-549b0400.pdf
Terinte, N., Manda, B. M. K., Taylor, J., Schuste, R. K. C., & Patel, M. K. (2014). Environmental assessment of coloured fabrics and opportunities for value creation: Spin-dyeing versus conventional dyeing of modal fabrics. *Journal of Cleaner Production, 72*, 127–138. https://doi.org/10.1016/j.jclepro.2014.02.002
Tischner, U. (2011). *System innovation for sustainability* (pp. 2–5). Greenleaf Publishing.
Visser, M., Gattol, V., & van der Helm, R. (2015). Communicating sustainable shoes to mainstream consumers: The impact of advertisement design on buying intention. *Sustainability (Switzerland), 7*(7), 8420–8436. https://doi.org/10.3390/su7078420
Wibowo, B. Y., Widodo, W., Widiyanto, & Setyawan, N. A. (2023). The sustainability of Lasem batik in terms of green marketing and brand quality. *Quality - Access to Success, 25*(198), 284. https://doi.org/10.47750/QAS/25.198.31

Wibowo, B. Y., Widodo, W., Widiyanto, & Setyawan, N. A. (2024). The Sustainability of Lasem batik in terms of green marketing and brand quality. *General Management, 25*(198), 284–291. https://doi.org/10.47750/QAS/25.198.31

Widjajanti, E., Regina T. P., & Utomo, M. P. (2011). Zeolite adsorption patterns on Azo Methyl Red and Methyl Orange Dyes. *Proceedings of the national seminar on research, education and application of MIPA,* K115-K122, Fakultas MIPA, Universitas Negeri Yogyakarta.

Widyasti, A. R., Lestari, A., Amri, K., Naufal, F., & Budiasih, K. S. (2017). Development of standardization of natural batik dye from secang bark (*Caesalpinia sappan* L.) using spectroscopic techniques. *Saintek Research Journal, 22*(1), 49–58. https://doi.org/10.21831/jps.v22i1.14850

Wongwilai, S., Phudetch, P., Saelek, P., Khuptawatin, A., Wongcharoensin, K., Chaitongrat, S., Vaiyavuth, R., & Jermsittiparsert, K. (2022). The role of innovative ideas in business sustainability: Evidence from textile industry. *Uncertain Supply Chain Management, 10*(1), 285–294. https://doi.org/10.5267/j.uscm.2021.8.011

Xia, Y., & Li-Ping Tang, T. (2011). Sustainability in supply chain management: Suggestions for the auto industry. *Management Decision, 49*(4), 495–512. https://doi.org/10.1108/00251741111126459

Yaacob, M. R., & Mat Zain, N. F. (2016). Environmental awareness and perception on environmental management of entrepreneurs in batik industry in Kelantan, Malaysia. *Journal of Tropical Resources and Sustainable Science (JTRSS), 4*(1), 54–60. https://doi.org/10.47253/jtrss.v4i1.574

Yernisa, Sa'id, E. G., & Syamsu, K. (2013). Application of natural powder coloring from areca nut extract (Areca catechu L.) to color transparent soap. *Journal of Agricultural Industrial Technology, 23*(3), 190–198. Retrieved from https://journal.ipb.ac.id/index.php/jurnaltin/article/view/7908

Yoshanti, G., & Dowaki, K. (2017). Batik life cycle assessment analysis (LCA) for improving batik small and medium enterprises (SMEs) sustainable production in Surakarta, Indonesia. *EcoProduction*, 997–1008. https://doi.org/10.1007/978-981-10-0471-1_68

Žurga, Z., & Tavčer, P. F. (2014). Apparel purchasing with consideration of eco-labels among Slovenian consumers. *Fibres and Textiles in Eastern Europe, 22*(5), 20–27. http://www.fibtex.lodz.pl/article1334.html

The Development and Challenges of Infant Diapers Ensuring Good Health with Sustainability Aspects: Landscape of UN Perspective

Abhijit Mukherjee

1 Introduction

Replacement of a newborn's nappy is now considered essential to baby care. Modern diapers have gone quite a long way in meeting babies' needs while adhering to social conventions, being a necessary component of human life. The stage thus set is that the book presents not only a detailed discussion of the basics and complexities of the diaper itself but also an examination of design elements and materials, and an overview of sustainability. Since then, diaper design has come a long way in accommodating technology and materials science and reflecting societal changes. For generations, traditional cloth diapers have been in use. Nowadays, selecting a single option is exceedingly challenging, as many alternatives range from conventional clothing to highly sophisticated disposables.

There are advantages and disadvantages to each style to accommodate the needs of both the babies and the caregivers. At its core, nappy design is about absorbing and retaining liquids, but it also includes waste control, skin health care and movement facilitation. Super-absorbent polymer (SAP), prevention of nappy rashes and other common skin disorders in newborns and the presence of harmful ingredients are all thoroughly investigated using scientific methods. Succeeding in this endeavour requires an expert mix of engineering and scientific knowledge. So, this chapter explores the ingredients and structure of diapers in great detail, illuminating the complex compositions that make them effective. However, serious concerns about sustainability come hand in hand with technological progress. The most recent developments in ecologically friendly diapers, an excellent replacement for

A. Mukherjee (✉)
Department of Fashion Technology, National Institute of Fashion Technology, Kolkata, West Bengal, India
e-mail: abhijit.mukherjee@nift.ac.in

S. S. Muthu (ed.), *Sustainable Textile and Apparel Chain Management*, Sustainable Textiles: Production, Processing, Manufacturing & Chemistry,
https://doi.org/10.1007/978-3-031-80240-9_7

traditional disposables, are also addressed in this chapter. These cutting-edge innovations may reduce the negative ecological impacts without sacrificing the performance and safety aspects of the product. Anticipating future diaper developments focusing on reusable diapers and other solutions, this brings together theoretical research integrated with real-world experiences for improving baby care solutions via innovation and sustainability.

In brief, this chapter overviews diaper developments aligning with the UN's SDGs, global trends with regional insights, holistic analysis and evaluation focusing on absorption, fit, skin protection, environmental impacts and future direction combating ecological challenges.

1.1 Diapers: Essential for Modern Infant Care

Infants must wear diapers to maintain cleanliness and comfort and to remain protected from infections and skin irritations, such as diaper rashes. Two types of diapers are commercially available: *disposable and cloth (reusable) diapers*. The advantages and disadvantages of each type of diaper must be considered while procuring baby care.

The plastic, tissues and absorbent gel, known as SAP, are widely used in disposable diapers for convenience. They may be disposed of after a single use, making them perfect for busy parents in this era. However, their frequent disposals lead to environmental hazards, and the recurring expenditures may be substantial. On the other hand, cloth diapers are made of different types of fabrics and offer an eco-friendly option. They are reusable, allowing parents to save money over time and save the baby from exposure to toxic chemicals that might be present in commercial disposable diapers. However, compared to disposables, they necessitate more effort and time for regular cleaning and maintenance.

The diaper size is essential for proper fit and preventing leakage and skin irritation. The diaper sizing requires the measurement of the user's waist and hip circumference. Diaper rash is a typical issue with baby diapers. This is generated due to moisture on the skin's intimate layers and friction between the skin and diaper layer. To prevent this development and growth of the rashes and maintain proper hygiene of the tender skin, changing the diaper frequently every couple of hours or more often, if required, is essential. Cleaning the bottom parts of the baby with warm water and mild soap is, however, advisable while changing the diaper every time. Baby wipes, free from fragrance and alcohol, can also be used, followed by patting the skin dry with a soft towel. Applying a diaper cream containing zinc oxide helps safeguard the top layer of skin from friction and humidity, preventing skin irritations.

For effective newborn baby care, diapers are essential to use. Diaper use helps keep the baby comfortable and promotes the infant's hygiene. Though disposable diapers are easier to use, cloth diapers are more environmentally friendly. So, to avoid diaper rashes, frequent changes of diapers along with choosing the correct size are also significant for effective baby care (Atherton, 2004; Blume-Peytavi et al., 2014; Blume-Peytavi & Kanti, 2018; Lund et al., 2001).

1.2 UN Sustainable Development Goals

There are seventeen well-defined sustainable development goals (SDG) to transform the world for better living, which are as follows (Fig. 1):

(a) *No poverty*: This aims to "*end poverty in all its forms everywhere*" by ensuring access to basic needs like health, education, and water and sanitation, especially for those living at extreme poverty levels. Poverty badly affects all countries, including the rich countries in the world.
(b) *Zero hunger*: This aim is to "*end hunger, achieve food security and improved nutrition and promote sustainable agriculture*," as hunger is detrimental to the productivity of people more prone to diseases.
(c) *Good health and well-being*: Aims to "*ensure healthy lives and promote well-being for all at all ages*" by providing access to safe and affordable medicines and vaccinations to all people under universal health coverage.
(d) *Quality education*: Aims to "*ensure inclusive and equitable quality education and promote lifelong learning opportunities for all*" as education encourages gender equality, empowers people to maintain healthy and sustainable lives and develops tolerance among people, ultimately building peaceful societies.

Fig. 1 Sustainable Development Goals (SDG) of the United Nations. (https://www.un.org/sustainabledevelopment/news/communications-material/)

(e) *Gender equality*: This aim is to "*achieve gender equality and empower all women and girls*" as gender inequality stagnates social progress. Gender equality and women's empowerment lead to a peaceful, prosperous, and sustainable world.
(f) *Clean water and sanitation*: This aim is to "*ensure the availability and sustainable management of water and sanitation for all*" as water scarcity not only leads to rising temperatures and global climate change but also falls under fundamental human rights, including sanitation and hygiene.
(g) *Affordable and clean energy*: This aim is to "*ensure access to affordable, reliable, sustainable, and modern energy for all*" as clean energy sources like solar, wind, etc. can encourage growth and help the environment.
(h) *Decent work and economic growth*: Aims to "*promote sustained, inclusive and sustainable economic growth, full and productive employment and decent work for all*" to reduce informal employment and gender pay inequality, especially for young people, and to promote a safe and secure work environment with sustainable economic growth.
(i) *Industry, innovation, and infrastructure*: This agenda aims *to "build resilient infrastructure, promote inclusive and sustainable industrialization, and foster innovation,"* especially in transport, irrigation, energy, and information and communication technology. These are crucial for sustainable development and the empowerment of communities in many countries.
(j) *Reduced inequalities*: Aims to "*reduce inequality within and among countries*" by ensuring equitable resource distribution, investments in education and skill development, implementation of social protection measures, combating discrimination, supporting marginalized groups and fostering international cooperation for fair trade.
(k) *Sustainable cities and communities*: This aims to "*make cities and human settlements inclusive, safe, resilient and sustainable*" by ensuring access to safe and affordable housing, upgrading slum settlements, investing in public transport, fostering urban governance and planning through participatory and inclusive ways and developing green spaces.
(l) *Responsible consumption and production*: Aims to "*ensure sustainable consumption and production patterns*" required to sustain current lifestyles through implementing sustainable procurement policies.
(m) *Climate action*: Aims to "*take urgent action to combat climate change and its impacts,*" which include weather pattern change, sea level rise, and frequent extreme weather events. This will also provoke mass migrations, leading to instability and wars.
(n) *Life below water*: Aims to "*conserve and sustainably use the oceans, seas and marine resources for sustainable development*" by allocating increased funds for ocean science, intensifying conservation efforts, and turning the tide on climate change to safeguard the planet's largest ecosystem.
(o) *Life on land*: Aims to "*protect, restore and promote sustainable use of terrestrial ecosystems, sustainably manage forests, combat desertification, halt and reverse land degradation, and halt biodiversity loss*" to minimize the loss of

biodiversity and habitats of nature that are an integral component of our shared heritage.

(p) *Peace, Justice and strong institutions*: Aim to "*promote peaceful and inclusive societies for sustainable development, provide access to justice for all and build effective, accountable and inclusive institutions at all levels*" by promising inclusive engagement at all times, combating corruption, and minimizing the flow of illicit armaments by promoting human rights and reinforcing the rule of law.

(q) *Partnerships for the goals*: Aims to "*strengthen the means of implementation and revitalize the global partnership* (between government, the private sector, and civil society) *for sustainable development*" through mobilizing both existing and additional resources by individuals and fulfilling official development assistance commitments by developed countries (https://www.un.org/en/exhibits/page/sdgs-17-goals-transform-world).

1.3 Current Status of Usage and Access of Diapers Across the World and UN

The infant diapers market is on an upward trajectory globally with an annual growth rate of about 4.43% (CAGR 2024–2029), driven by the rising population, accompanied by the increasing urban population, the growing birth rate and the need for convenient and comfortable products for babies; increased disposable incomes; heightened awareness of infant health and an increase in the number of working women; the increasing rate of literacy, determining the living standards and consumption patterns of families. The Baby Diapers Market is anticipated to attain USD 59.10 billion by 2029, with an estimated USD 47.58 billion in 2024. Each individual contributes US$7.34 per capita revenue and is expected to use 1.1 kg of diapers per individual, on average, in 2024. Moreover, there is an estimated volume growth of 3.1% in 2025.

Dominated by disposable diapers for convenience and hygiene, the baby diapers market faces increased competition from substitute products, such as cloth diapers, which offer a more sustainable and eco-friendly alternative. In recent years, there has been a growing trend toward using cloth diapers, driven by the growing consciousness of the ecological impacts of disposable diapers and the desire for a more sustainable solution. Both conventional and organic products are in high demand, with the latter experiencing a significant increase due to health and ecological concerns. With ongoing advancements in the fabric used in diapers, competitors' market share has increased due to diapers' special user-friendly features and absorption capabilities. The distribution is extensive, encompassing both online and offline channels. The growing popularity of e-commerce platforms for baby care products is also expected to increase the sales of baby diapers through online channels, owing to their adaptability and variety.

The market in North America and Europe is characterized by a demand for premium products, which is supported by substantial purchasing power. The Asia Pacific region is anticipated to undergo the most significant growth due to its vast demographics, urbanization, and improved living standards. China alone documents the birth of about 18 million infants annually in the Asia-Pacific region. The country is responsible for the majority of the global infant diaper market. Implementing the third-child regulations in China has significantly augmented the demand for infant diapers. In 2019, Rwanda, Madagascar, Nepal, Burundi, Tanzania, Zimbabwe, Cambodia, Vietnam, South Sudan and Iceland experienced the highest participation of women's labour force worldwide.

Consequently, it is anticipated that product demand will increase in those countries in the years ahead. The global market for infant diapers has surged in the Asia Pacific region, particularly in recent years. The high rate of births in African countries also facilitates the adoption of products. In order to meet the requirements of contemporary parents, the competitive landscape is characterized by an unwavering commitment to innovation that prioritizes sustainability and user-friendliness (https://www.statista.com/outlook/cmo/tissue-hygiene-paper/baby-diapers/worldwide) (https://www.mordorintelligence.com/industry-reports/baby-diapers-market) (https://www.statista.com/outlook/cmo/tissue-hygiene-paper/baby-diapers/worldwide#analyst-opinion) (https://www.grandviewresearch.com/industry-analysis/baby-diapers-market).

Compared to other regions of the globe, the purchasing capacity of customers in North America and Europe is comparatively high. This increased purchasing power enables them to allocate more funds toward premium, high-quality infant products. It is anticipated that countries with a high proportion of women in the workforce will also experience a significant increase in demand for infant products due to their increased purchasing power. However, with the introduction of The End Diaper Need Act (EDNA) of 2021 by Sen. Tammy Duckworth in February 2021, low-income families could afford diapers for their babies and toddlers (https://www.grandviewresearch.com/industry-analysis/baby-diapers-market).

The challenges for single-use/disposable diapers regarding indiscriminate disposal of materials in environmentally sensitive areas, such as waterways and rangelands, resulted from the lack of adequate waste collection systems in rural and peri-urban areas, the lack of adequate solutions or technologies for appropriately disposing of, the low propensity for recycling and degradation (according to the World Wildlife Fund, taking over 500 years to decompose) (https://www.undp.org/eswatini/blog/exploring-sustainable-solutions-single-use-diapers).

From the above discussions, it is clear that sustainable infant diapers will offer the best alternative to overcome all challenges in the future world. They align with UN Sustainable Development Goals such as "*Good Health and Well-being, Responsible Consumption and Production, Life on Land, Sustainable Cities and Communities, and Climate Action*."

1.4 Main Function of the Diaper

Diapers are essential in maintaining infants' overall well-being, comfort and hygiene. They act as a protective barrier between a baby's delicate skin and bodily waste (primarily urine and faces) by absorbing moisture promptly and containing waste discharged during defecation. Diapers prevent skin irritation, diaper rash and discomfort caused by prolonged exposure to wetness.

Diapers are essential in promoting the infants' hygiene and cleanliness, ensuring the intimate skin remains dry and fresh throughout the day and night, enhancing the baby's comfort and minimizing the risk of infections and other health issues. By simplifying the process of diaper changes and waste disposal, diapers provide convenience for parents and caregivers, facilitating more efficient and effective management of infant hygiene. Overall, the users cannot ignore the importance of diapers in infant care, as they contribute significantly to health and comfort, enhancing the overall quality of life for the babies and their caregivers. About 4600–4800 diapers are predicted to be consumed within the first 3 years of a child's life (Dey et al., 2016).

The main functions of a diaper are to provide solutions for the baby's skin against excreta such as urine and faces, which are as follows:

(a) *Prevents waste containment and leakages*: Diapers are designed to retain waste, prevent leakage, and avoid the risk of messes all over clothes, bedding, and other surfaces.
(b) *Moisture absorption and liquid retention for skin protection*: Diapers are designed for quick absorption and sustained retention of liquids, keeping the baby's skin dry and preventing discomfort and irritation of the outermost skin from prolonged exposure to friction and humidity.
(c) *Odour control*: Diapers include materials that help neutralize and retain odours from urine and faeces, resulting in a more pleasant environment for infants and caregivers.
(d) *Facilitation of care*: Diapers help parents and caregivers maintain their baby's hygiene, providing a simple and sanitary means for waste disposal and diaper changes.

1.5 Basic Requirements for Baby Diaper

The desirable properties and requirements for diapering materials for various stages of the babies are as follows:

(A) *New-born stage (0–5 months)*: Enhanced tactile properties, such as extra softness for skin intimate layers, adequate absorbency for soft stool and urine retention, and good leakage protection to maintain dryness of the tender skin.

(B) *Determined explorer stage (6–17 months)*: Good absorbency and adequate stretch (elastic recovery) with a comfortable fit (especially thin and narrow between the legs) that is adaptive to free body movements.
(C) *Capable learner stage (18–23 months and above)*: Pant-shaped diapers capable of being put on/taken off by the babies themselves are of paramount demand (Wiesemann & Adam, 2011).

So, the typical requirements in all stages of development may be included as follows:

1. Superior absorbency with thinnest and flexible substrates, offering maximum skin dryness and comfort with a reduced substrate mass.
2. Leakage protection through all sides of the diaper.
3. Skin protection from all possible infections from microbes (viruses and bacteria).
4. Wearing comfort with better softness and improved breathability.
5. Sustainable products are compatible with reducing overall adverse environmental effects over the product's lifespan by minimizing energy consumption for manufacturing, transport and distribution, maintenance, and generation of landfill wastes.

1.6 Types of Diaper

In the modern world of parenthood, diapering stands as a bastion of innovation and choice. From the disposable diaper, revered for its unparalleled convenience and swift cleanup, to the eco-suitable cloth diaper, highly revered for its reusable nature and environmental mindfulness, the spectrum of options is vast.

Few diapers, fortified with extra layers of protection, emerge as steadfast allies for uninterrupted sleep, fostering an environment of comfort and reassurance for both parents and babies. For the discerning few with sensitivities or environmental convictions, organic diapers stand as a beacon of purity, which offers a chemical-free sanctuary for delicate skin and a sustainable choice for the conscientious consumer. From a market perspective, the diapering landscape reflects innovation and consumer-centricity, with each variant of a distinct set of needs and preferences ensuring a multi-dimensional demand for modern parenthood.

Therefore, in continuation of the above discussion, there are several types of diapers commercially available in the market, with their distinctive features and purposes:

(A) *Disposable diaper*: A disposable baby diaper is a single-use absorbent undergarment made of a combination of components such as polypropylene, polyethylene and a super-absorbent polymer, along with minor amounts of tapes, elastics and adhesive materials for effectively collecting and containing a baby's waste effectively. It has a moisture-wicking upper layer with absorbent softcore and attached fasteners that users can adjust to provide a snug fit. These are quick to change and simple to use. Disposable diapers help caregivers to keep their babies comfortable and clean. Keeping the baby's skin dry is a barrier against leakage and prevents skin irritations or rashes. Still, some types can also be responsible for diaper rashes due to fungal or bacterial growth, even with exposure to toxic substances present in the diaper. Disposing of used disposable diapers properly is critical to preserve cleanliness and sanitation. The advantages and disadvantages of disposable diapers are discussed below:

(a) *Advantages*

(i) Very convenient and easy to use.
(ii) Highly absorbent.
(iii) Ideal for travel as they are lightweight and easy to dispose of.
(iv) Numerous disposable diapers are equipped with elastic leg cuffs and leak shields, which are safety features.

(b) *Disadvantages*

(i) Typically made with chemicals instead of fabrics.
(ii) Become very costly over time.
(iii) They contribute significantly to generating landfill wastes, which take a long time to decompose, leading to serious environmental concerns.
(iv) Increases the risk of diaper rash due to the materials and chemicals used in their construction.
(v) May lead to a dependency on them, making the transition to potty training more challenging.

Disposable diapers offer convenience and absorbency, but they come with significant long-term costs, both environmentally and financially. While they require minimal upfront investment and are easy to use, their contribution to landfill waste is substantial.

(B) *Reusable cloth diapers*: A cloth diaper is a reusable diaper usually made of cotton, hemp, bamboo, and microfiber fibres. It is multi-sized, multi-shaped and simple to use and reuse. Every cloth diaper has an exterior waterproof layer and an inner absorbent layer. The construction of these layers makes a difference; while some absorbent inserts are integrated, others are detachable.

1. *Advantages*

(i) Generates less waste.
(ii) Substantial cost savings over time.
(iii) Made from skin-friendly materials.
(iv) It is comparatively sustainable and eco-friendly.

2. *Disadvantages*

(i) Consumes more energy and water.
(ii) Significant cleaning and laundry time are required.

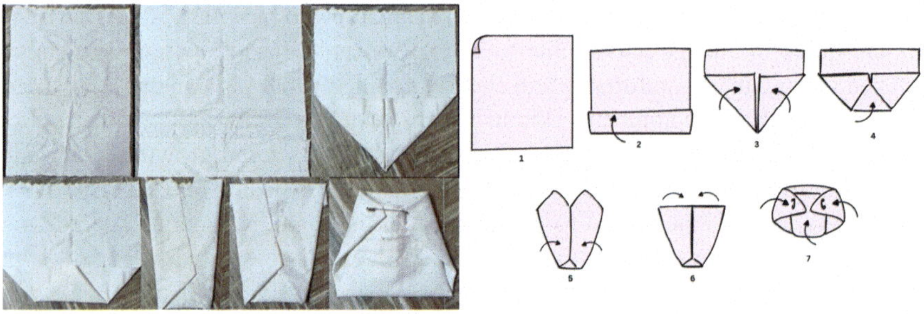

Fig. 2 Flat diaper. (adapted and modified from https://www.pinterest.com/pin/233342824426333296/)

(iii) Might be less day-care-friendly.
(iv) It has less liquid holding capacity.

Various Types of Reusable Cloth Diapers

The following is a compilation of the seven most prevalent types of cloth diapers:

1. *Flat diapers*: A flat diaper is a singled-layered piece of fabric that is usually made of cotton, which is capable of being folded in a variety of ways, resulting in a multi-layered, diaper-like shape fitting the baby. To secure this flat cloth reusable diaper, one needs safety pins or clips and a diaper cover to keep moisture inside the diaper (Fig. 2).

Flat cloth diapers offer an affordable and versatile option for parents embracing this cloth diapering option. They also allow customization of absorbency by sub-layering, making them adaptable to a baby's changing needs. However, achieving a snug fit requires mastering folding techniques. While flat cloth modules have multiple uses beyond diapering, their reliance on traditional folding methods may pose a challenge for caregivers unfamiliar with cloth diapering practices.

2. *Pre-fold diapers*: Pre-fold diapers are very similar to flat diapers but are pre-folded into three panels, keeping the middle panel the most absorbent. Pins or safety pins are also necessary for securing the pre-folds, and a diaper cover is suggested to prevent leakage (Fig. 3).

Pre-fold cloth diapers offer an economical and practical choice for parents interested in cloth diapering systems. They come pre-layered, saving time for adjustment compared to flat diapers. However, they still require folding into an exact diapering shape and securing with pins or clips. Like flat diapers, they also need a separate cover for leak protection. Despite their affordability and adjustability, pre-fold diapers may also pose challenges for caregivers unfamiliar with cloth diapering techniques.

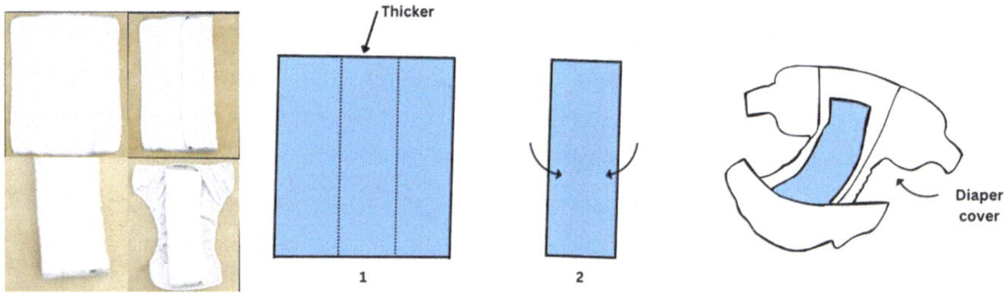

Fig. 3 Pre-fold diaper. (adapted and modified from https://thiswestcoastmommy.com/6-ways-to-fold-a-prefold-diaper/; https://images.app.goo.gl/wVxSGDDMKQ7W1wDEA)

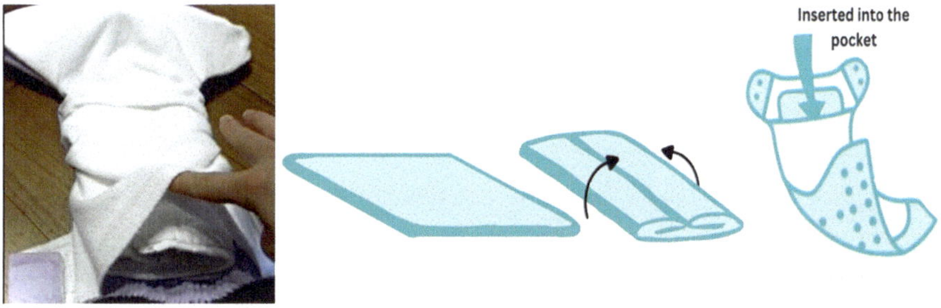

Fig. 4 Pocket diaper. (adapted and modified from https://images.app.goo.gl/wVxSGDDMKQ7W1wDEA; https://images.app.goo.gl/fGg8cyNUFdGQr25b6)

3. *Pocket cloth diaper*: Pocket diapers are shaped like disposable diapers; these cloth diapers have elasticated waist and leg apertures for easy fastening with efficient waste containment for easy adjustments, usually with snaps or velcro fitted with this module. These diapers' main feature is a pocket inside with a provision for stuffing the absorbent inserts. Numerous cloth diaper inserts, made with various disposable or reusable materials such as microfiber, cotton or even hemp fibres, are available in the market (Fig. 4).

Pocket cloth diapers offer convenience without folding requirements, making them ideal for parents or caregivers. The absorbent insert is separate and detachable, making washing and drying easier, though drying takes longer. Disposable inserts may be employed with specific pocket diapers during travel.

4. *All-in-one cloth diapers (AIO)*: All-in-one (AIO) cloth diapers are one-piece diapers that do not require any extra accessories or separate inserts. The waterproof cover is stitched between the absorbent layers, eliminating the need for additional steps while using it for the infant (Fig. 5).

The built-in fasteners of all-in-one cloth diapers make them quite convenient for caregivers. However, they have a longer drying time and less customizable absorption, making them less suitable for overnight use (https://www.madeformums.com/reviews/nappies-what-types-are-there/).

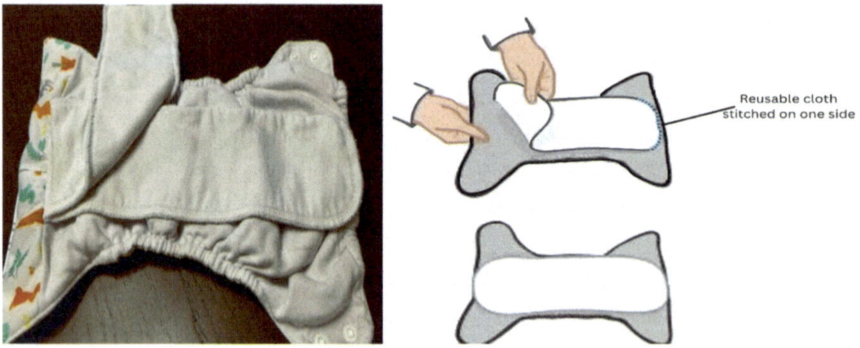

Fig. 5 All-in-one diaper. (adapted and modified from https://images.app.goo.gl/wVxSGDDMKQ7W1wDEA; https://images.app.goo.gl/fGg8cyNUFdGQr25b6)

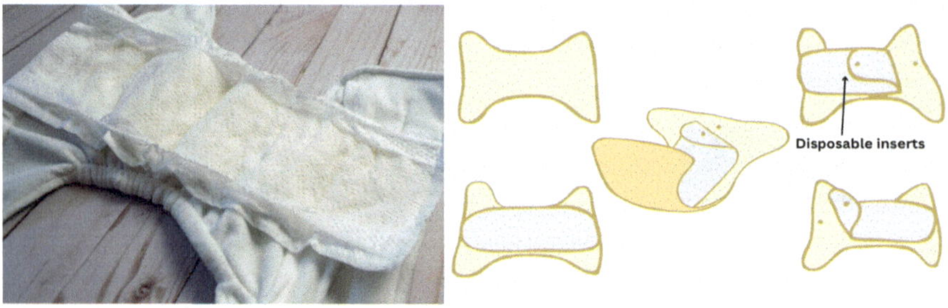

Fig. 6 Hybrid diaper. (adapted and modified from https://www.simplymombailey.com/a-look-at-the-new-pampers-pure-hybrid/; https://images.app.goo.gl/wVxSGDDMKQ7W1wDEA; https://images.app.goo.gl/fGg8cyNUFdGQr25b6)

5. *Hybrid cloth diapers*: Hybrid cloth diapers were designed to make cloth diapering convenient and sustainable by combining reusable covers with disposable inserts. They reduce the hassle of laundering by allowing for easy disposal of soiled inserts, making them ideal for busy parents or can be used while travelling (Fig. 6).

However, their cost is higher due to the need to purchase disposable inserts on an ongoing basis. Hybrids decrease waste compared to traditional disposables but still generate disposable waste to a lesser extent. However, they are economical in the long run compared to disposables.

6. *Fitted cloth diapers*: These require an outer waterproof diaper cover to prevent leakage. However, the absorbent inner is the actual diaper. They are relatively simple to place on a baby. Fitted diapers offer the baby excellent absorbency coverage, which works well at night (Fig. 7).

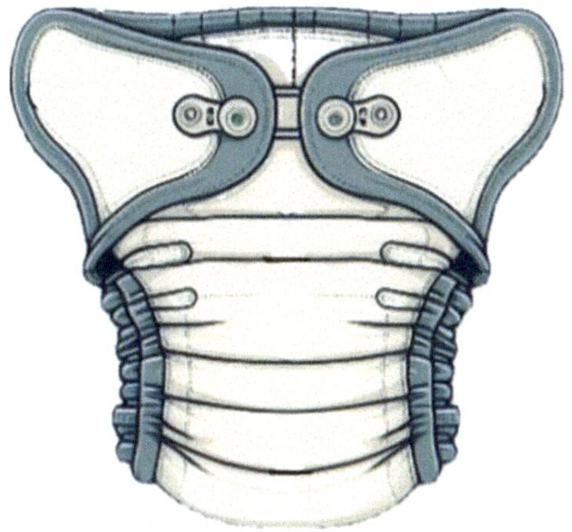

Fig. 7 Fitted diaper

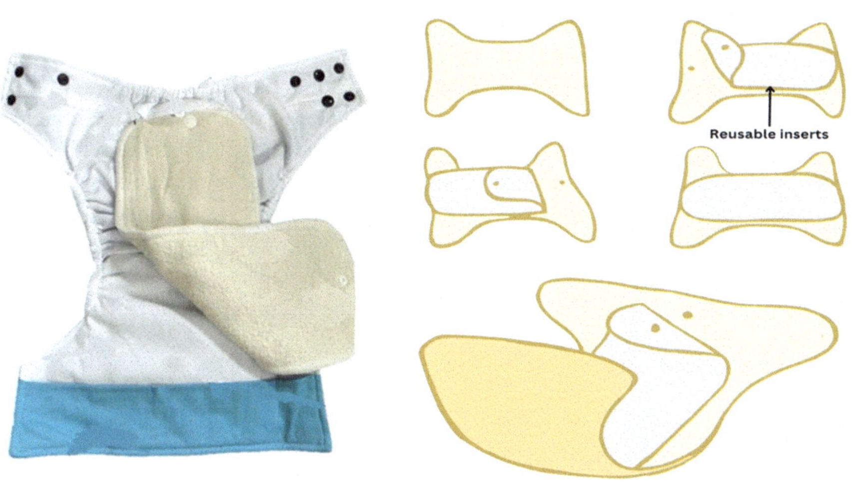

Fig. 8 All-in-two diaper. (adapted and modified from https://images.app.goo.gl/wVxSGDDMKQ7W1wDEA; https://images.app.goo.gl/fGg8cyNUFdGQr25b6; https://www.allaboutclothdiapers.com/peek-inside-a-pocket-all-in-two-cloth-diaper/)

Fitted cloth diapers do not require pins, snaps or Velcro secures the inner layers. However, they can be more expensive than the basic options available, which require a two-step diapering process and may take longer to dry due to multiple layers.

7. *All-in-two cloth diapers (AI2)*: All-in-two (or AI2) diapers have two parts: *inserts and covers*. They resemble pocket diapers in many ways, except the insert is often fastened within the diaper rather than placed inside a pocket. This feature is convenient as it allows one to easily remove and put in the wet insert without changing the complete diaper (Fig. 8).

Some AI2s can be used with disposable inserts, but these are more costly than other cloth diapers. Additionally, the outer covers may be used again, as wet inserts can be removed and replaced.

All-in-two cloth diapers offer convenience but may have drawbacks, such as messy insert removal and limited absorbency customization with snap attachments (Aishwariya & Priyanka, 2020).

A brief comparison among all types of reusable diapers is shown in Table 1.

1.7 Comparison Between Disposable and Cloth Diapers

The comparative analysis between cloth reusable diapers and disposable diapers is shown in Table 2.

From the above comparison table, it is clear that cloth diapers offer numerous benefits, including environmental friendliness, potential cost savings in the long run and a chemical-free option for a baby's dedicated skin. They also foster a closer connection to the diapering process and may even aid in potty training by promoting awareness of wetness. On the other hand, disposable diapers provide convenience, especially while travelling or during hectic times, and they often come with better moisture-wicking properties to keep babies feeling dry and comfortable. They can also be a practical choice for families with limited time or access to laundry facilities. Selecting disposable diapers means sacrificing the customization and sustainability of cloth diapers. Despite their initial appeal, the cumulative expense of disposable diapers over time and their negative impacts on the environment and sustainability should be carefully considered.

1.8 Parts of a Diaper

Disposable diapers made from everyday materials have a long history of safer use in various standard consumer products. The primary components are cellulose, polypropylene, polyethylene, super-absorbent polymer, tapes, elastics and adhesive materials. The further development of lighter, thinner, and more effective diapers resulted from the advancement of these materials. The most recent advancements are creating biodegradable brands and exploring renewable resources for raw materials to reduce the environmental impacts of disposable diapers. Similarly, cotton, bamboo, hemp and other microfiber and accessories such as buttons and snap fasters used for reusable diapers are discussed in Sect. 1.6 (B).

Modern disposable diapers underwent several stages of improvement in terms of component materials and construction techniques, resulting in improved performance, fit and wear comfort.

Table 1 Comparison among all reusable diapering systems

Subjects	Flat diapers	Pre-fold diapers	Pocket cloth diaper	All-in-one cloth diapers	Hybrid cloth diapers	Fitted cloth diapers	All-in-two cloth diapers
Absorbent	Flat folded sheet	Tri-folded sheet	Padded inserts	Sewn-in	Disposable inserts	Whole diaper	Padded inserts
Requires diaper waterproof cover	Yes	Yes	No	No	Yes	Yes	No
Easy laundering	Easier	Easier	Easy	Easy	Super easy	Easier	Easy
Easy changing	Easier	Easier	Easy	Easy	Easy	Easy	Easy
Cost	The most inexpensive cloth diapering option	Relatively low in cost	It is costlier than flats and pre-folds but economical in the long run	It is costlier than flats and pre-folds but economical in the long run	More expensive, owing to the requirements for frequent repurchase of disposable inserts	It is costlier than flats and pre-folds but economical in the long run	It is costlier than flats and pre-folds but economical in the long run
Advantages	Very affordable, customizable, quick drying time	Very affordable, customizable, pre-layered	Very easy fit, customizable absorbency	Easy fit, one-piece module	Easy to use, quite convenient for cleaning up, little laundering required	Reusable outer layer, comprehensive absorbency, and easy fit	Easy fit, reusable outer layer
Disadvantages	Requires folding, requires separate fasteners and covers	Requires separate fasteners and covers	More expensive than flats and pre-folds require stuffing of absorbent inserts	Longer drying time, less customizable absorbency	Expensive, generates landfill waste	Extended drying time, most require purchasing multiple sizes	Numerous demands more than a single piece, absorbency that is fewer possibilities for customization

https://www.charliebanana.com/blog/types-of-cloth-diapers-cloth-diapering-guide/

Table 2 Features of cloth and disposable diaper

	Disposable diapers	Cloth reusable diaper
Convenience	Very convenient, no washing required	Require washing after each use
Skin sensitivity	It is made up of synthetic chemical-based materials, which can cause skin irritations due to toxicity	It is made of different fibres/fabrics free from chemicals and safe for baby skin
Environmental impact	Increase landfill waste and require large amounts of chemical-based raw material production	Reduces landfill waste but may also increase water and energy use for regular maintenance
Sustainability	Not environmentally sustainable	Environmental friendly, especially if made of organic and sustainable materials
Leakage management	Rely on an absorbent core made of SAP (sodium polyacrylate) to prevent leakages	Leg and waist elastics may contain messes, preventing leakage
Absorbency	Excellent absorbency keeps baby skin dry	Depending on the type of cloth diaper used, it may require more frequent changes
Customizations	Limited customization options are available	Various styles and materials are available for diverse customization
Cost	It can be more expensive over time	High initial investment but cost-effective in the long run
Fit-related concerns	It requires purchasing multiple sizes with the baby's growing body, and sometimes, a wrong-sized diaper can cause severe skin infections	One size may be expanded to fit a growing child

https://www.charliebanana.com/blog/cloth-diapers-vs-disposable-pros-and-cons/

Parts of a Diaper

Diapers are made up of the following primary parts (Fig. 9):

(a) *Top sheet*: This is a specialized hydrophilic component that is both soft and dry, intended to preserve moisture retention and reduce the risk of skin irritation by efficiently channelling fluids faster to the diaper core. Some top sheets include moisturizers, which can safeguard the tender skin from irritation and over-hydration (Ersser et al., 2005; Forslind et al., 1997; White, 2003).

(b) *Diaper core*: The middle core layer, also known as the absorbent core, where urine and faeces are absorbed and reserved, is usually made of soft materials or SAP (in the case of disposable diapers, which are made up of different combinations of PP, cellulose pulp and super-absorbent polymers). This layer absorbs and traps moisture within its structure, keeping it away from the baby's skin and preventing skin rashes.

- *Absorbent core*: In the 1980s, researchers developed super-absorbent polymers, known as SAP, made of sodium polyacrylate, that can absorb liquids many times their weight. The interior layer of the diaper is the absorbent core

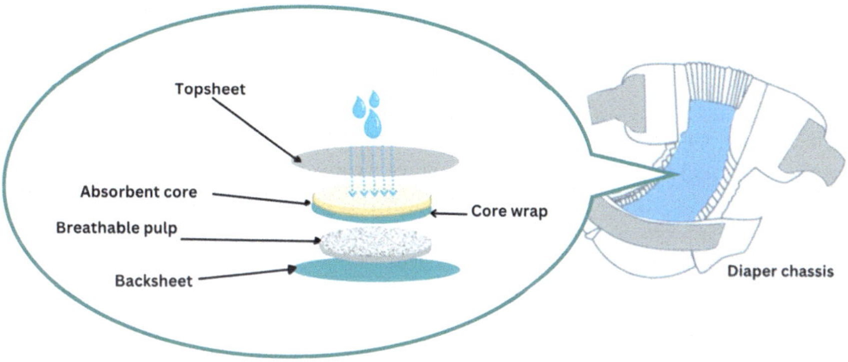

Fig. 9 Anatomy of the diaper

and generally comprises a combination of polyacrylate granules and cellulose fluff pulp. The cellulose component quickly absorbs and transfers urine fluid to the super-absorbent polyacrylate material for immediate trapping. This super-absorbent material can retain moisture up to 30 times the weight of the substrate polymer, which, in turn, facilitates a significant reduction of diaper weight and volume due to the decrease in consumption of absorbent, leading to reduced adverse environmental impact (Buchholz & Graham, 1998).

(c) *Backsheet*: It is the waterproof outer layer of the diaper, restrains liquid leakage through it and prevents the ruining of the baby's clothes and the surrounding environment. The polyethylene backrest can be made of a PE structure, allowing only water vapour and air (not the liquid) to pass through. This reduces inside moisture, keeping the skin dry. It is lightweight and mostly breathable.
(d) *Diaper chesis*: The diaper chesis includes the back sheet and additional features such as fastener-secured securing systems, leg cuffs, tapes and stretch side panels, which ensure the diaper fits well and holds all the parts together.

- *Fit and comfort features*: Diapers are designed to fit comfortably against the body. A diaper's three most crucial components are the stretch side panels, the fastening mechanism and the elastics.

The stretch side panels must be wide enough for the diaper to form a good seal on the baby's torso. The fastening mechanism should facilitate seamless adjustments around the waist without being too tight. To prevent leakage, the elastics on either side of the diaper must be of suitable size (Ozen et al., 2016).

1.9 Material Used for Manufacturing Disposable Diapers

An average disposable diaper weighs between 40 and 51 g and is composed predominantly of cellulose, polypropylene, polyethylene and a super-absorbent

Table 3 Materials and their uses in disposable diapers

Material used	Uses
Super-absorbent polymer (SAP)	Absorbent core
Cellulose pulp substrate	Absorbent core and core tissue
Polypropylene (PP)	Used in the top sheet, the diaper part next to the baby's skin
Polyethylene (PE)	Used in the back sheet, the outer cover, tapes and fastening systems
Elastics	Used for leg opening and waist, which improves fit
Adhesives	Used for bonding various other materials together all through the diapers

polymer, with minor quantities of adhesive materials, elastics, and tapes. Due to advancements in all these materials, diapers have become lighter, more effective in performance and less impactful on the environment.

The materials utilized in disposable diapers have undergone numerous modifications and constructions to improve their performance, fit, and skin comfort. These advancements have significantly minimized the environmental impact of disposable diapers in recent years.

Materials Used in Diapers

The material used in disposable diapers is shown in Table 3.

(a) *Super-absorbent polymer*: Super-absorbent polymers, also called SAP, are primarily used as water absorbents in diapers. These super-absorbent polymers are reported to absorb about 30 times or more liquids of their weight. It is also called SAM (super-absorbent material). It is typically a granular substance similar to table salt in appearance. This improves the liquid retention capacity, which, in turn, facilitates the disposable diaper to be much thinner (about 20%) with limited usage of fluff pulp and elevated performance (Dey et al., 2014).
(b) *Cellulose fluff pulp*: This is used to construct the absorbent pad in the diaper core to absorb liquid. It offers integrity and absorbing capacity to the diaper. In addition, the pulp has a high fibre ratio to the substrate's weight. The fibre is of lower coarseness with a shorter fibre length.
(c) *Polypropylene (PP)*: Polypropylene nonwoven fabric is used as the back sheet, giving the diaper a comfortable shape and preventing liquids from leaking. This keeps the baby's skin dry even after allowing liquid to pass through, making it an excellent hydrophobic material with a higher wicking property.
(d) *Polyethylene (PE)*: It is mainly used in sanitary products and diapers for its higher water absorption (specially, amidoxime-based polyethylene) capacity. Compared to other nonwoven materials, it has been gaining acceptance due to its physical properties, such as lightweight, reusability, toughness and flexibility.

    - *Top Sheet surface add-on lotions*: Several topical lotions are added to the nonwoven top sheet, such as aloe vera, vitamin E, almond oil, vitamin D, neem, etc. Tertiary ammonia or silver salt compounds have been used as antibacterial lotions recently. However, many paediatricians are against the use of it for various reasons.

(e) *Elastics*: These, usually made of polyester foam, synthetic rubber or spandex, are crucial in making a diaper effective in performance and with improved fit to the body.
(f) *Adhesives*: Adhesives prevent leakage by securely holding the absorbent fibres or super-absorbent powder inside the diaper core. They are also used to keep the many strands of elastic together around the legs and waist areas of the diaper.

1.10 Fibres Used for the Manufacturing of Cloth Diapers

Cloth diapers are mostly made from organic materials or fibres such as cotton, hemp, bamboo, fleece and microfiber. These materials are selected for properties such as softness, absorbency, breathability and sustainable nature. They are free from harmful chemicals, making them gentle on the baby's skin and eco-friendly. The fibres/fabric substrates most often used in diapers are as follows:

(a) *Bamboo fibre* is a regenerated cellulosic fibre made from raw materials extracted from bamboo plants. Bamboo is a member of the plants that grow at the quickest rate on the planet, making it very renewable. The properties of bamboo are anti-allergic, antibacterial, antistatic, absorbent, breathable, thermoregulation, durable and soft. It is highly absorbent since it can absorb up to four times more liquid than cotton. This regenerated fibre is also known as bamboo viscose. Bamboo kun, which is naturally present in bamboo, is reported to be responsible for its antimicrobial properties (Renukadevi & Poornima, 2007; Shanmugasundaram & Gowda Mahendra, 2010).
(b) *Cotton* is the most widely cultivated natural fibre on this planet. It has long been used in diaper manufacturing.

    Organic cotton is cultivated without synthetic pesticides, herbicides or fertilizers and processed under strict ecological conditions. Its ability to remain dry against the skin is one reason for its popularity in diaper top sheets; however, cotton also gained customer preference in diapers because of its soft and comfortable appeal.
(c) *The fleece* structure is very soft to the touch. It does not absorb liquid; it only helps the liquid pass through, which makes the surface dry. The baby wearing a pocket diaper feels dry as the moisture is taken into the multilayer of very absorbent bamboo inserts. *Polyester fleece* is available in various weights and variants and performs several functions. The thinnest substrate of polyester flees is often used inside the diaper, which acts as a dry layer, and a thicker substrate may be used in the breathable cover for diapers.
(d) *Polyester* is a synthetic material made of petroleum extracts. The word "polyester" may be replaced by "microfiber" and occasionally by "zorb" in diapering

industries. These are pretty absorbent, but they wear out over time, losing their ability to absorb with continued wear. Microfiber should never be used directly against a baby's skin as it can be irritating, so it is used in pocket diapers. It may be challenging to get clean after each washing, being very absorbent if the substrate is made up of microscopic structures and, therefore, may develop odours over time afterwards.

(e) *Hemp* has natural antifungal and antibacterial properties, making it an effective cloth diaper fabric that can lessen the chances of skin irritations and allergies. Hemp is the strongest fibre and four times more absorbent than other absorbents like cotton, being very hygroscopic in nature. It is also a sustainable fibre. It is very thin but incapable of promptly absorbing water like cotton or polyester. So, for this reason, it is often blended with other absorbent fibres.

(f) *The waterproof layers* of a cloth diaper are frequently composed of substances such as PUL (polyurethane laminate) and TPU (thermoplastic polyurethane), which prevent leakage.

2 Construction of a Diaper

(a) *Construction of the absorbent pad*: It is a crucial phase in developing an absorbent diaper. The absorbent pad is developed by drawing a long conveyor belt through a chamber filled with pressurized nozzles. The bottom side of the conveyor belt pierces, and the fibres are drawn downward by a vacuum, forming a flat pad. The surface of the pad was covered with an absorbent material after it had been constructed. This results in the blockage of moisture retained in the external layer, which cannot be disseminated to the centre. This may hold water against the skin and cause discomfort. Then, different sprayers were applied to the pad in different layers of polymer and fibre. After that, the fibre is drawn to the centre, forming the bottom pad. Then the polymer is included in the mixture to form a combined layer of fibre and polymer. The pad then proceeds along the conveyor belt to get levelled near the forming chamber exit. Then, the uniform thickness of the pad is ensured by removing (by rollers) some of the fibre from the top of the pad.

(b) *Forming the nonwoven fabric for the inner and outer layers*: After developing the absorbent pad, it is time to make the diaper. In the case of cloth diapers, they are made of different fabrics and a cloth-based absorbent pad. However, in disposable diapers, plastic resins form nonwoven sheets through a meltdown process, and the guiding substrate is made through wide rollers that cut the fabric to the appropriate width. This process is for both top and bottom sheets.

(c) *Component assembly*: The final step in the diaper manufacturing process is to assemble every part of the absorbent pad, the top sheet and the bottom sheet, whether a disposable diaper or a cloth diaper (Kabiri, 2003).

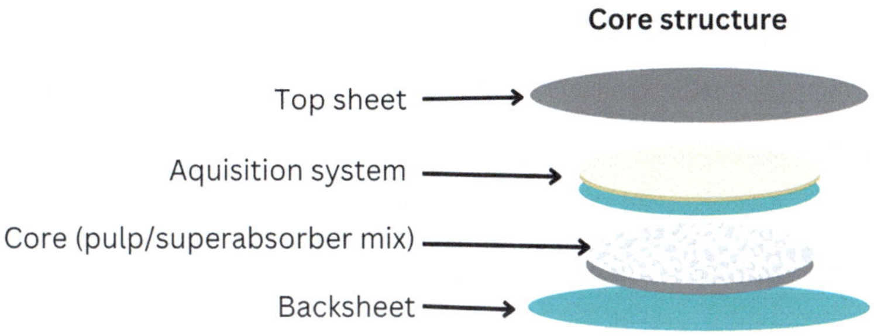

Fig. 10 Anatomy of the core structure of disposable diaper. (adapted and modified from Dey et al., 2016)

(a) A disposable diaper consists of three layers, namely (Fig. 10) absorbent top layer, which is made of nonwoven fabric, an absorbent pad in the core and a waterproof bottom layer. About 1000 diaper designs and technology have been patented in the last few years. The diapers, made of cloth, were produced in mass in past times. Materials such as wood pulp fluff, cellulose wadding, fluff cellulose or cotton substrates with comparatively less absorbing capacity were used in early disposable diapers (Erasala et al., 2011). The cellulose is extracted from pine trees and made into fluff pulp, which can draw liquid by capillary action. Following 1980, several attempts were made to modify the structure to enhance the absorbing capacity, with more biodegradable materials being used, the mass of the product being reduced and the product being better fitted to the wearer. Recently, absorption polymers have been introduced in diapers to enhance their efficacy. Polyacrylate [super-absorbent polymer (SAP) gel] was introduced along with fluff pulp, with a liquid retention capacity of up to 30 times its weight. This also minimizes the thickness of the diaper, making it more comfortable for the wearer (child). Leading disposables provide an acquisition layer and a back sheet that help to maintain the product's performance without compromising skin health (Counts et al., 2014).

(b) Diaper dermatitis is of an irritant contact nature and predominantly appears in children of 6–12 months of age (Kuwayama et al., 1992). Factors responsible for diaper dermatitis include excessive wetness, friction and elevated pH levels with high enzymatic activities, which are developed due to the interaction of faeces and urine, compromising the skin's epidermal barrier function. This interaction between these initial factors leads to skin inflammation, triggering the diaper rash (Stamatas & Tierney, 2014). Skin wetness and maceration give rise to the susceptibility to friction between the skin and the diaper substrate, causing physical damage to the skin. Faecal ureases cause the breakdown of urea into ammonia, escalating the skin's pH. Excessive skin irritation produces erythema, which arises from the enhanced activity of faecal enzymes, proteases, ureases and lipases. These enzymes also enhance the susceptibility to additional potential irritants (Andersen et al., 1994; Berg, 1987; Berg, 1988; Berg et al., 1986; Buckingham & Berg, 1986; Erasala et al., 2007; Van Gysel, 2016).

(c) Introducing smart diapers has recently brought innovation to the disposable diaper market. A small strip fixed in front of the diaper contains four small coloured squares that respond differently when coming into contact with specific proteins, water content or bacteria. These squares will change their colours if abnormal reactions crop up, signalling caregivers for diaper replacement. The advantage of these smart diapers is that they offer less frequency in diaper changes, relax anxious parents and make frequent verifications for replacements. A few producers have also developed an application that notifies the parents as soon as the baby wets. A sensor is integrated into the diaper, which senses moisture and alerts the parent through a smartphone application. These smart diapers are more expensive (about 30% or more) than conventional diapers but are still in the prototype phase for further development (Odio & Thaman, 2014; Srinivas & Sandipan, 2016).

(d) In the diaper structure, the urine is first promptly absorbed by the top sheet, which has a soft and porous nonwoven surface. The absorbed liquid should be transferred downwards to keep the baby's skin dry. Urine is then promptly absorbed at the ADL (acquisition and distribution layer) and transmitted to the absorbent (core) layer. The polyester fibres lie in the ADL to ensure the urine is evenly spread and, consequently, faster transmission to the absorbent substrate along the fibre length assembly. The back sheet, usually composed of non-bioactive hydrophobic coatings, functions as a microporous barrier beneath the core layer. This microporous layer prevents urine leakage from dropping through the diaper clothes (Odio & Friendlander, 2000; Clark-Greuel et al., 2014. Counts et al., 2017).

In the absorbent core layer, two different raw materials- pulp and SAP[wood pulp and super-absorbent polymers] are uniformly spread between two nonwoven surfaces and enveloped. The SAP used in the core retains liquid to about 30 times its weight, which is about 1.5 times its weight for wood pulps. The pulp (containing high-density and evenly distributed short fibres) ensures the even distribution of urine and a little absorption. The pulp layers' raw material is typically extracted from specific trees cultivated in the North American region. The cavities in the fibre structure facilitate the passage of liquid.

Super-absorbent polymer (sodium polyacrylate), used in fine-grained structures such as table salt, increases liquid retention in the diaper, ensuring its higher performance using a thinner substrate of less mass. In contact with water, the sodium polyacrylate (SAP) generates negatively charged carboxyl ions and diverges the solution. It is soluble in water, and sodium ions absorb water. Afterwards, this polymer gets transformed into a gel in the aqueous solution. The three-dimensional structure (high molecular density) of gel is thus formed by being interconnected by cross-linking. Acrylate is capable of retaining hydrogen (H-O-H) in water as a result of polarity forces (Fig. 11). The super-absorbent properties of SAP are induced by electrolytes (0.9% mineral electrolyte) present in the urine. Electrolytes minimize polarity. This results in the development

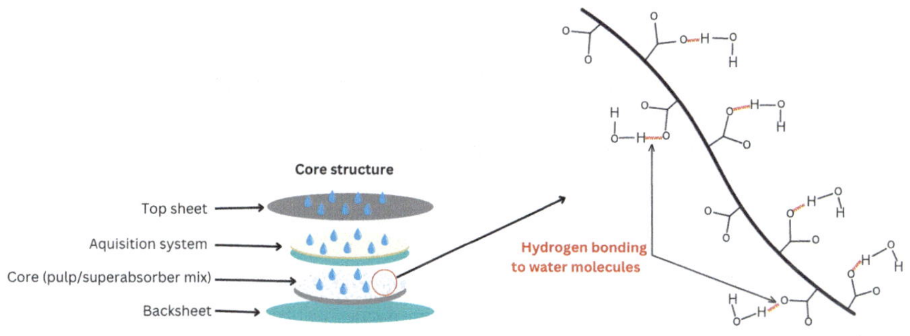

Fig. 11 Absorption and reaction mechanism of SAP (sodium polyacrylate) in aqueous environment. (adapted and modified from Kotz et al., 2009)

of SAP's super-absorbent capacity, which is required for fluid retention. Therefore, the efficacy of the diaper with SAP should never be verified with pure water; it can be tested with a 0.9% saline solution (Campbell et al., 1987; Helmes et al., 2014; Lane et al., 1990; Odio & Friendlander, 2000).

The formulation (mixture and distribution) of pulp and SAP in the diaper is crucial so the mixture can be condensed in a particular portion rather than distributed evenly. So, it is understood that these two should be used more meticulously. In other words, the absorbent layer must be formed in three dimensions. Accordingly, a new method is adopted to deposit most of the SAP material around the middle zone of the diaper. Popular brands of diapers also confirm this understanding. As a result of this advancement, the reduction of usage of pulp (cellulose) and chemicals (SAP) is thus achieved, leading to a decrease in disposable waste, further ensuring the restoration of a sustainable green world (Gencturk et al., 2021).

Several clinical investigations revealed a reduction of skin dermatitis by about 50% by using SAP and a significant improvement of the humidity and breathability of the skin along with maintaining the skin's pH under the diaper area (Clark-Greuel et al., 2014; Odio & Friendlander, 2000). It has also been reported that faecal contamination for mothers and caregivers is being restricted (Campbell et al.,1988a, b; Odio et al., 2013).

(e) Applying a breathable outer layer made up of microporous materials ensures the water vapour only passes through diaper layers, restricting liquid seepage from the inside. This limits the severity of infection caused by diaper dermatitis (Candida albicans) by about 38–50% by maintaining skin moisture at the desired level (Akin et al., 2001; Odio & Thaman, 2014).

(f) The petrolatum-based formulation (petrolatum or zinc oxide) is applied on the outer cover in patches without compromising the absorbency of the diaper substrates. The continuous light coating of the skin by the above formulation restricts the proximity of the irritant to the skin under the diapering area, maintaining smoothness between two adjacent surfaces (Odio et al., 2000; Odio & Thaman, 2014). Even the safe use of non-sensitizing pigments instead of disperse dyes, ingrained with diapering material, minimizes the leakage of excess liquor from the diaper. This results from the insoluble nature of high molecular weight pigment molecules in aqueous solution (Evans et al., 2014). The price range of this diaper varies depending on the diaper brands and the manufacturing technology used. Even introducing diapers well impregnated with barrier creams helps minimize diaper dermatitis, maintaining the safe relative humidity level of skin under the diapering area (Srinivas & Sandipan, 2016).

(g) The flushable diaper is an ultra-thin substrate with a suitable size and shape, offering comfort to the wearer. This is made of bio-degradable microfiber (patented), easy to flush off into the toilets after use, fragmented into small particles during flashing agitation without blocking the pipes and ultimately deposited to the water treatment plant. The thin cover stock (like tissue paper) gets exposed, releases the contents inside and diffuses in the water completely. In the end, the flashed-off contents, being organic in nature, are entirely disintegrated by bacteria present in the septic system (Espinosa-Valdemar et al., 2014; Ferronato et al., 2020; Khoo et al., 2019; Lörcks, 1998).

2.1 The General Performance Testing for Baby Diapers

The following parameters are considered to be evaluated for the performance of the baby diapers:

1. *Liquid strike-through test*: This test measures the time it takes for the liquid to be transferred to the inner layer from the upper layer of the diaper substrate. A drop of test liquid (solution) is placed on the test sample to measure the time it takes to reach the inner layer of the diapering material (Cottenden et al., 1998).
2. *Absorption capacity*: The purpose of the diaper lies primarily in the absorption of the liquid (urine) without rewetting the surface immediately after it is dispersed to the inside layers for retention. The total absorption capacity is determined by the EDANA method by dipping the test sample into a known test solution for a standard time and measuring the weight difference from the initial one before testing (Hardy, 2009).
3. *Acquisition time and rewet under load*: These are significant properties for baby diapers. Acquisition time is measured as the time taken by the diapering material for complete absorption of a known amount of test liquid under a pre-determined load simulated to the standard weight of a baby (approximately 2.5 Kg). It is essential as the problem of leakage and fit under the simulating condition can be evaluated, along with the issue of odour neutralization of the urine (Cottenden et al., 1997, 2006).

Rewetting under load is the ability of the diapering material to rewet the surface under the simulated load of the baby (approximately 2.5 Kg). In other words, the amount of liquid absorbed and is squeezed out under load subsequently while the baby sits down. It is very much related to the skin dryness achieved with diapering material (Wiesemann & Adam, 2011).

2.2 Different Innovative Applications/Developments in Diaper

(a) *Development of disposable baby diapers with microencapsulated natural fragrance finishes*: Fragrant finishing in textiles imparts pleasant scents or fragrances to fabric surfaces. This technique is achieved by applying specialized chemicals or additives, and these textiles serve both aesthetic and functional purposes. This type of finishing is done using the microencapsulation technique. This technique can significantly ensure the controlled release rate of fragrance compounds and other essential oils as desired.

The baby diapers were manufactured using a variety of fibrous compositions, including polypropylene spun bonds on the top sheet, a highly absorbent core and pure bamboo spun lace supported by polypropylene melt blown at the bottom sheet. The bottom sheet, being non-porous and hydrophobic, helps the outer layer to remain dry. Pure bamboo spun lace nonwoven treated with fragrant oils of Citronella and Marjoram, applied by microencapsulation technique, was placed to diffuse fragrant smell when pressure is applied while sitting or moving. Super-absorbent polymer (SAP) was then superimposed on the pure bamboo spun lace microencapsulated material. Pure polypropylene spun bonds are used as top sheets to transport the urine through it, keeping the baby's tender skin dry. Elastic and fasteners are fitted at the sides of the diaper, followed by sealing all sides with an ultrasonic welding machine. This fragrant finish would provide a pleasant smell and enhance the freshness of the product (Annapoorani et al., 2016).

(b) *Development of baby diapers made of bamboo and organic cotton fibre*: The research was conducted to develop a baby diaper from four different fibrous compositions: *pure bamboo, pure organic cotton, bamboo-organic cotton blends (70:30) and bamboo-organic cotton blends (50:50)*. The tests for antibacterial activities were carried out on the prepared diaper samples against the bacteria S. aureus and E. coli. The reduction of antibacterial activity on pure bamboo fibre shows 100% and 98.75% against E. coli and S. aureus, respectively. However, organic cotton fibre shows zero reduction towards the bacteria. The study also highlighted various parameters such as liquid strike-through time, absorption capacity and acquisition time, which exhibited high efficacy on the bamboo fibre (Shanmugasundaram & Gowda Mahendra, 2010).

(c) *Development of diapers made of regenerated carboxylated cotton fibre*: Cellulose has been proven to be a significant functional part of the diaper structure. The fibre's stiffness, quality and fundamental reliability are imparted by its presence. Due to the elimination of lignin, hemicelluloses and other substances, the higher cellulosic fibre strands increase the effective surface area, which, in turn, enhances mechanical interlocking by providing a better interfacial bond between the fibres and a larger area available for interaction with the fibre lattice structures. These filaments have the potential to be used as reinforcement for producing composites since they are biodegradable, lightweight and environmentally friendly (Brown & Malcolm, 2007). Diapers composed of such regenerated cellulosic fibres show better absorbency than natural cotton fibres.

They will act similarly to very spongy polymers (SAP) and the consumption system of liquids after the cellulose strands are altered. The positive sodium ions in ionic polymers, the cellulose derivatives, neutralize the negative carboxylate groups, making them electrically nonpartisan. The sodium particles become wet and lose their attraction to the carboxylate groups when the polymer reacts with water; this allows the sodium ions to move freely inside the polymer structure. This process leads to the formation of an osmotic pressure inside the gel. Extreme swelling occurs with deionizer water; little or no swelling occurs with liquids with a high salt concentration because of low osmotic influence (Kabiri, 2003; Shabrin & Palaniswamy, 2019).

(d) *Development of bio-degradable baby diapers*: An effort is made to develop a biodegradable infant diaper by substituting the polypropylene nonwoven with a biodegradable viscose nonwoven that has explicitly been completed to prevent pollution and ensure that the product satisfies its functional requirements. The results suggest that the absorption capacity is exceptional, with a value of 3.8 g/g fulfilling the primary function of the diapering material. The specimen requires 4.2 s to transmit the fluid in the liquid strike-through test, even though it takes more than other commercial items. The wetback data indicates a meagre value of 1.03 g/g. The risk of diaper damage is minimal, as the retention capacity (3 kpa) under load is only 3.8 g/g. Therefore, these biodegradable diapers not only satisfy the functional requirements but also appear more environmentally friendly (Sasikumar et al., 2014).

(e) *Development of organic cloth diapers with disposable coconut coir fibre lining*: The research was conducted to utilize the coconut coir fibre for organic cloth diapers to address sustainable concerns and improve baby hygiene. Coconut coir is naturally abundant and eco-friendly, exhibiting high water absorption, retention and aeration properties, making it ideal for diaper linings. The study compares diaper performance and concludes that coir fibre linings outperform absorption capacity and liquid strike-through time for commercial disposable diaper applications. Additionally, the process of sourcing coir fibre involves sun-drying coconut husks and grinding them into powder, exploring its sustainable nature. This research underscores the potential of coir fibre in diaper production, offering a viable solution for eco-conscious parents while ensuring baby comfort and skin health (Mishra & Basu, 2020; Paramanandham et al., 2014; Thilakavathi & Ramakrishnan, 2021).

(f) *A smart wearable for diaper moisture quantification and notification*: Wet diapers make babies feel uncomfortable and can create health issues like diaper rash. Diaper rash can be restricted by replacing it immediately after the baby leaves bodily wastes (urine and faeces). This inspires the development of a wearable gadget that automatically alerts the parental or caregiver's smartphone when an infant urinates.

Using a decision tree and a midpoint-based k-nearest neighbour (KNN) hybrid classification algorithm, the proposed wearable detects urination by sensing the temperature increase on the exterior surface of the diaper. The device is attached to the diaper's external layer and is compact, low-powered and cost-effective. It is equipped with a reusable electronic device. The device can be integrated into any disposable diaper without requiring any modifications to the diaper manufacturing process. This system can facilitate the treatment of diseases that necessitate precise records of urination. Subsequently, a hardware device and a smartphone application prototype were created and evaluated (Atherton, 2004; Gupta & Skinner, 2004; Khan, 2019).

2.3 Skin Rashes

Diaper rashes are another term for diaper dermatitis, an acute, episodic inflammatory condition caused in the skin under a diapered area. It is rarely severe from a medical point of view but can cause discomfort for infants and anxiety for caregivers. Several studies estimate the rate of diaper dermatitis, which varies significantly from 16% to 65%, and this effect remains for 2–4 days (Stamatas et al., 2014; Adam, 2008; Ward et al., 2000; Jordan et al., 1986; Adalat et al., 2007; Blume-Peytavi & Kanti, 2018).

The prevalence of rash and its causes are still unknown, which has prompted studies to determine unsafe substances in diapers. A few studies have been done on heavy metals in infant disposable diapers to determine the concentrations of heavy metals and other harmful materials. However, more studies are still needed on what causes diaper rashes, even if there have been positive changes in diaper properties.

(A) *Types of Diaper Dermatitis or Diaper Rashes*

The following are the different types of diaper dermatitis:

(i) *Irritant dermatitis*: It is the most common and is due to the arising of high alkaline pH of the skin, moisture, mixture of stools and urine or even rubbing between diaper and skin (Atherton, 2016; Bender et al., 2017).
(ii) *Infectious dermatitis*: These diaper rashes can again be distinguished into two types.
   (a) *Yeast infection*: A common cause of diaper rash is yeast infection, which is generated by an overgrowth of fungi (Candida albicans) found naturally in the digestive tract.
   (b) *Bacterial infection*: Diaper rash can also result from bacterial infection. This is also called impetigo. Bacteria like *Staphylococcus aureas* and

Escherichia coli (E. coli) may cause diaper rash or worsen existing ones (Šikić et al., 2018).

(iii) *Inflammatory dermatitis*: Occasionally, babies with sensitive skin may have an allergic reaction to some specific ingredient in diapers, wipes or even skin creams. The frequently occurring allergens include dyes or elastics in diapers and fragrances or preservatives in diaper wipes or creams (Yu et al., 2017).

Determining harmful substances like formaldehyde and heavy metals is essential as it directly interacts with the baby's delicate skin. Most diapers usually have dyes, and some may contain heavy metals, which cause allergic reactions in the baby's skin. In an investigation by Alberta et al. (2005), the dyes in diapers were found responsible for diaper rash in babies, and it manifested only in the area of skin which was in intimate contact with the dyed portion of the diaper. The symptoms were found insignificant with the use of dye-free diapers (Giusti et al., 2003).

Diaper rash may be caused by leaving wet or soiled diapers on the skin for a long time. Tight-fitting diapers that rub against the skin can also lead to a rash. A bacterial or yeast infection may develop and expand to the surrounding skin. The region covered by a diaper is at risk because of trapped moisture at elevated temperatures, making it ideal for bacteria and yeast growth. These rashes can be seen in the creases of the skin.

(B) *Symptoms*

(a) Inflamed skin under the diaper area, such as thighs, genitals and buttocks.
(b) Itchy and swollen skin under the diaper area.
(c) Sore skin under the diaper-wrapped area.
(d) Discomfort and irritation, especially during diaper changes.

Impact on Diaper Choices

A critical observation of five concerns, which are skin care, infection control, other health-related issues, environmental and safety concerns and time and cost considerations, indicates the following:

(a) Super-absorbent paper diapers diminish the incidence and intensity of diaper dermatitis and handle the spread of infection in caregiving surroundings. However, sometimes, they can cause severe issues for sensitive baby skin.
(b) Cloth diapers have different effects on the skin environment, and none is found to be exclusively superior.
(c) The retail value of disposable, reusable, and laundered diapers may be comparable, although home-laundered diapers are not very expensive and could be one of the best solutions.

Again, there is no agreement on whether disposable or cloth diapers are the best for minimizing the risk of diaper rashes. However, most paediatricians believe that disposable diapers prevent inflammation and diaper rash. This prevention keeps the baby's bottom dry, but it has an environmental impact as it takes longer for decomposition.

Even parents who use cloth diapers can minimize the risk of infection by reducing the time the diapers spend on the skin with urine and faeces. However, if the diapers are changed frequently, diaper rashes can be prevented in both types of diapers (Nyamukamba et al., 2023).

2.4 Sustainability Concerns

Diaper belongs to a vital group of products used in the parenting journey. Apart from their usability, disposable diapers are usually thrown away after one-time use, resulting in waste management problems. However, cloth diapers only require laundering (Khoo, 2019).

As previously mentioned, disposable baby diapers and other products like adult incontinence products and menstrual pads have the largest market share in sales of absorbent hygiene products (Newswire, 2021).

This resulted in the consumption of basic materials to manufacture disposable diapers, leading to increased waste generation. Over time, public awareness of environmental issues and sustainability related to diapers has increased.

As a result, different solutions have been developed nowadays to mitigate the environmental burdens associated with diaper production, usage, and disposal, such as introducing eco-design practices, eco-friendly waste treatment, and recycling processes for soiled diapers (Cordella et al., 2015).

Diapers are a type of absorbent hygiene product (AHP) that come in various sizes, shapes and materials (Ng et al., 2013).

Research has shown that a baby uses about 7000 diapers, on average, during the diapering period before the toilet training starts, and about 95% of diapers used are disposable types (EDANA, 2008). There is increasing awareness and concern over the ecological issues associated with diapers, including air, water and ground pollution; resource utilization; and waste production. This has driven researchers to consider greener design options and technologies to reduce such burdens (Alhogbi, 2017). However, environmental concerns associated with disposable and reusable diapers are different.

These concerns are addressed below:

(A) *Environmental Impacts of Disposable Diapers*

- (i) The waste in disposable diapers has the potential to seep into groundwater, posing a risk to the environment.
- (ii) Made primarily of different types of plastics based on petroleum extracts that are not biodegradable.
- (iii) It contains numerous harmful chemicals, such as SAP (super absorbent and other polymer), which are released into the environment, and some are very carcinogenic.

(iv) These disposable diapers end up in landfills and take hundreds of years to decompose. In addition, they also emit methane, a greenhouse gas, causing ozone layer depletion and facilitating global warming.
(v) Releases ozone-depleting chlorofluorocarbons that have a negative impact on the environment, causing global warming.

(B) *Environmental Impacts of Reusable Cloth Diapers*

(i) Much water is required to wash cloth diapers for regular maintenance.
(ii) Regular washing requires more energy (electricity), and water consumption leads to more resource utilization and unsustainable practices.

Biodegradable disposable diapers: Biodegradable disposable diapers are a particular category of single-use diapers composed of various materials designed to decompose relatively faster in the environment (Alhogbi, 2017). Biodegradable single-use diapers can be a better option. Usually, these diapers are made of natural fibres from plants or animals (Asim et al., 2015; Mukhopadhyay & Vinay, 2008; Patiño-Masó et al., 2019; Ramamoorthy et al., 2015; Shanmugasundaram & Gowda, 2011). Materials used in biodegradable diapers are silk, wool, pineapple, sisal, bamboo, etc., but these diapers can be more expensive (Mendoza et al., 2019).

2.5 Challenges and Future Scopes

1. To build awareness among consumers to accept that the thinner air-laid core can manage better urinary flow than other thicker fibre/fabric substrates.
2. In the context of skin protection, the following factors play an essential role in the development and growth of diaper dermatitis, favoured by the increase in humidity, pH changes (resulting from the development of ammonia by the interaction of urine and ureases from faecal microbes), challenges with minimizing mechanical friction on the wet skin and chemical changes with skin microorganisms promoting colonization by the virus-like *Candida albican* and the bacteria-like *Staphylococcus aureus*. (Berg, 1988; Runeman, 2008).
3. The development of super-absorbent polymer (SAP), enhanced ability for absorption and retention of liquid (under slight mechanical pressure) in the range of about 30 times the weight of the substrate, has evolved into a cutting-edge invention in diapering technology facilitating the production of ultra-thin diapers by the producers. The amount of application of cellulose pulp has been reduced by 50% with a high concentration of super-absorbent polymer in the core, resulting in thinner diaper products. Further, numerous research studies and applications of innovative diaper designs have been formulated using various combinations and forms of composites, SAP and other fibre substrates. As SAP material used in disposable diapers is not environmentally friendly, producers are still searching for biodegradable super-absorbent fibres used to produce diapers with reduced manufacturing costs and simplified construction techniques. Several

claims on biodegradability by different popular brands have been marketed, ultimately failing to meet environmental concerns (Buchholz & Graham, 1998; Buchholz & Peppas, 1994; Pytlik et al., 2000).
4. Airlaid pulp fabric is used for baby diapers (swim diapers), offering smaller pore sizes in the substrate for better distribution of moisture, rewetting properties and suitability to be used as core layer-substrates, compromising quick and higher absorption by the acquisition layers (Knowlson, 2001) (http://www.nonwoven.co.uk/reports/NTC%202002%20contents.html).
5. Bi-component or fibrillated fibres, nano fibres and fibre modification have the potential for a good replacement, with the conventional use of fibres/fabric substrates being a good filter for viruses and bacteria (Edwards et al., 2006).
6. The average diaper weight has been reduced by about 40% in the last two decades, followed by a significant reduction in packaging material, as raw material production is found to be the prime contributor to negative environmental impacts in the form of waste generation (especially for disposable diapers); requirement of energy for laundering and drying and consumption of groundwater and detergent for cleaning (EDANA, 2007).
7. Several epidemiological studies have shown that infants using diapers have some significant contribution to environmental contamination of microbes, especially in day-care settings. The disposable diapers, readily soiled by use and wrapped tightly in containers (packets), lead to the additional contamination of the faeces till safely disposed of (Van et al.,1991a, b; Campbell et al., 1988a, b).
8. In serious environmental and social responsibility concerns, all consumers and manufacturers should realize the relative environmental impacts of diapering materials in both the reusable and disposable options. Several LCA studies conducted in recent years around the world have reported environmental impacts in the form of resource depletion, acidification, ozone layer depletion, climate change, freshwater, aquatic toxicity, human contamination and eutrophication.

References

Adalat, S., Wal, L. D., & Goodyear, H. (2007). Diaper dermatitis—Frequency and contributory factors in hospital attending children. *Pediatric Dermatology, 5*, 483–488.

Adam, R. (2008). Skin care in the diaper area. *Pediatric Dermatology, 25*, 427–433.

Aishwariya, S., & Priyanka, P. (2020). A review on convenience and pollution caused by baby diapers. *Science & Technology Development Journal, 23*(3), 694–707. https://doi.org/10.32508/stdj.v23i3.2399

Akin, F., Spraker, M., Aly, R., Leyden, J., Raynor, W., & Landin, W. (2001). Effects of breathable disposable diapers: Reduced prevalence of Candida and common diaper dermatitis. *Pediatric Dermatology, 18*, 282–290.

Alberta, L., Sweeney, S. M., & Wiss, K. (2005). Diaper dye dermatitis. *Pediatrics, 116*, 450–452. https://doi.org/10.1542/peds.2004-2066

Alhogbi, B. G. (2017). Environmental issues and waste management for disposable and reusable diapers. *Environmental Science and Pollution Research, 24*(1), 1–10.

Andersen, P. H., Bucher, A. P., Saeed, I., et al. (1994). Faecal enzymes: In vivo human skin irritation. *Contact Dermatitis, 30*, 152–158.

Annapoorani, S. G., Nivedha, R. B., & Yamuna, D. S. (2016). Development of disposable diaper with microencapsulated natural fragrance finish. *International Journal of Applied Research, 2*(7), 685–687.

Asim, M., et al. (2015). Life cycle assessment of cloth vs. disposable diapers. *Journal of Cleaner Production, 99*, 197–205.

Atherton, D. J. (2004). A review of the pathophysiology, prevention and treatment of irritant diaper dermatitis. *Current Medical Research and Opinion, 20*(5), 645–649. Available from: https://www.madeformums.com/reviews/nappies-what-types-are-there/

Atherton, D. J. (2016). Understanding irritant napkin dermatitis. *International Journal of Dermatology, 55*, 7–9. https://doi.org/10.1111/ijd.13334

Bender, J. K., Faergemann, J., & Sköld, M. (2017). Skin health connected to the use of absorbent hygiene products: A review. *Dermatologic Therapy, 7*, 319–330. https://doi.org/10.1007/s13555-017-0189-y

Berg, R. W. (1987). Etiological factors in diaper dermatitis: A model for development of improved diapers. *Pediatrician, 14*, s27–s33.

Berg, R. W. (1988). Etiology and pathophysiology of diaper dermatitis. *Advances in Dermatology, 3*, 75–98.

Berg, R. W., Buckingham, K. W., & Stewart, R. L. (1986). Etiologic factors in diaper dermatitis: The role of urine. *Pediatric Dermatology, 3*, 102–106.

Blume-Peytavi, U., & Kanti, V. (2018). Prevention and treatment of diaper dermatitis. *Pediatric Dermatology, 35*(S1), s19–s23. https://doi.org/10.1111/pde.13495

Blume-Peytavi, U., Hauser, M., Lunnemann, L., et al. (2014). Prevention of diaper dermatitis in infants—A literature review. *Pediatric Dermatology, 31*, 413–429.

Brown, R. M., & Malcolm, R. (2007). In I. M. Saxena & M. Inder (Eds.), *Cellulose: molecular and structural biology: selected articles on the synthesis, structure, and applications of cellulose*. Springer. ISBN 9781402053801. OCLC 187314758.

Buchholz, F. L., & Graham, A. T. (1998). *Modern superabsorbent polymer technology*. Wiley.

Buchholz, F. L., & Peppas, N. A. (1994). *Superabsorbent polymers: Science and technology*. American Chemical Society.

Buckingham, K. W., & Berg, R. W. (1986). Etiologic factors in diaper dermatitis: The role of feces. *Pediatric Dermatology, 3*, 107–112.

Campbell, R. L., Seymour, J. L., Stone, L. C., & Milligan, M. C. (1987). Clinical studies with disposable diapers containing absorbent gelling materials: Evaluation of effects on infant skin condition. *Journal of the American Academy of Dermatology, 17*, 978–987. https://doi.org/10.1016/S0190-9622(87)70287-4

Campbell, R. L., Bartlet, A. V., Sarbaugh, F. C., & Pickering, L. K. (1988a). Effects of diaper types on diaper dermatitis associated with diarrhea and antibiotic use in children in day-care centers. *Pediatric Dermatology, 5*(2), 83–87.

Campbell, R. L., Bartlett, A. V., Sarbaugh, F. C., & Pickering, L. K. (1988b). Effects of diaper types on diaper dermatitis associated with diarrhea and antibiotic use in children in day-care centres. *Pediatric Dermatology, 5*, 83–87.

Clark-Greuel, J. N., Helmes, C. T., Lawrence, A., Odio, M., & White, J. C. (2014). Setting the record straight on diaper rash and disposable diapers. *Clinical Pediatrics (Phila), 53*(9 supple), 23S–26S.

Cordella, M., Bauer, I., Lehmann, A., Schulz, M., & Wolf, O. (2015). Eco-design and waste management strategies for disposable diapers. *International Journal of Life Cycle Assessment, 20*(4), 476–487.

Cottenden, A. M., Dean, G. E., & Brooks, R. J. (1997). Predicting the leakage performance of small body-worn disposable incontinence pads using laboratory test. *Medical Engineering & Physics, 19*(6), 556.

Cottenden, A. M., Thornburn, P. H., Dean, G. E., & Fader, M. J. (1998). Wet comfort of small disposable incontinence pads. *Textile Research Journal, 68*(7), 479.

Cottenden, et al. (2006). A critical investigation of ISO 11948-2 and ISO 11948-1 for predicting the leakage performance of small disposable incontinence pads for lightly incontinent women. *Medical Engineering & Physics, 28*, 42–48.

Counts, J. L., Helmes, C. T., Kenneally, D., & Otts, D. R. (2014). Modern disposable diaper construction: Innovations in performance help maintain healthy diapered skin. *Clinical Pediatrics (Phila), 53*(suppl 9), 10S–13S.

Counts, J., Weisbrod, A., & Yin, S. (2017). Common diaper ingredient questions: Modern disposable diaper materials are safe and extensively tested. *Clinical Paediatrics, 56*(5 supple), 23S–27S. https://doi.org/10.1177/0009922817706998

Dey, S., Helmes, C. T., White, J. C., & Zhou, S. (2014). Safety of disposable diaper materials: Extensive evaluations validate use. *Clinical Pediatrics (Phila), 53*(9), supple: 17S-9S.

Dey, S., Purdon, M., Kirsch, T., Helbich, H. M., Kerr, K., Li, L., & Zhou, S. (2016). Exposure factor considerations for safety evaluation of modern disposable diapers. *Regulatory Toxicology and Pharmacology, 81*, 183–193. https://doi.org/10.1016/j.yrtph.2016.08.017

EDANA. (2007). *Sustainability report 2007–2008: Absorbent hygiene products*. EDANA.

Edwards, S. L., Russell, S. J., Ingham, E., Mathews, J. B., & Mitchell, W. (2006). Nonwoven scaffolds of improved design for the tissue engineering of the anterior cruciate ligament. In S. C. Anand, J. F. Kennedy, M. Miraftab, & S. Rajendran (Eds.), *Medical textiles and biomaterials for healthcare* (pp. 355–365). Woodhead Publishing Ltd.

Erasala, G. N., Merlay, I., & Romain, C. (2007). Evolution of disposable diapers and reduction of diaper dermatitis. *Archives of Pediatrics, 14*, 495–500.

Erasala, G. N., Romain, C., & Merlay, I. (2011). Diaper area and disposable diapers. *Current Problems in Dermatology, 40*, 83–90.

Ersser, S. J., Getliffe, K., Voegeli, D., & Regan, S. (2005). A critical review of the inter-relationship between skin vulnerability and urinary incontinence and related nursing intervention. *International Journal of Nursing Studies, 42*, 823–835.

Espinosa-Valdemar, R. M., Sotelo-Navarro, P. X., Quecholac-Pina, X., et al. (2014). Biological recycling of used baby diapers in a small scale composting system. *Resources, Conservation and Recycling, 87*, 153–157. https://doi.org/10.1016/j.resconrec.2014.03.015

European Disposables and Nonwovens Association (EDANA). (2008). Sustainability Report: Baby diapers and incontinence products, prepared by The absorbent hygiene products manufacturers committee of EDANA. ISBN 2-930159-52-9.

Evans, E. B., Helmes, C. T., Kirsch, T., & Ruble, K. M. (2014). Colors in disposable diapers: Addressing myths. *Clinical Pediatrics (Phila), 53*(9), supple: 20S–2S.

Ferronato, N., Nova, P. M. L., & Torretta, V. (2020). Assessment of used baby diapers composting in Bolivia. *Sustainability, 12*(12), 5055. https://doi.org/10.3390/su12125055

Forslind, B., Engström, S., Engblom, J., & Lars, N. (1997). Novel approach to the understanding of human skin barrier function. *Journal of Dermatological Science, 14*, 115–125.

Gencturk, U., Ozgec, M., & Kalcoglu, A. (2021). Researching on baby diaper performance with 3D absorbent layer. In *The Eurasia Proceedings of Science, Technology, Engineering & Mathematics (EPSTEM)* (Vol. 12, pp. 95–105) (ISSN: 2602-3199).

Giusti, F., Massone, F., Bertoni, L., Pellacani, G., & Seidenari, S. (2003). Contact sensitization to disperse dyes in children. *Pediatric Dermatology, 20*, 393–397. https://doi.org/10.1046/j.1525-1470.2003.20504.x

Gupta, A., & Skinner, A. (2004). Management of diaper dermatitis. *International Journal of Dermatology, 43*(11), 830–834.

Hardy, P. (2009). *Baby diaper absorbent cores: New tests for effective core design*. Nonvovens Industry. Available: www.nonwovens_industry.com

Helmes, C. T., O'Connor, R., Sawyer, L., & Young, S. (2014). Disposable diaper absorbency: Improvements via advanced designs. *Clinical Pediatrics (Phila), 53*(9), supple: 14S–6S.

http://www.nonwoven.co.uk/reports/NTC%202002%20contents.html. Accessed 17 July 2024 at 3:31 PM.

https://images.app.goo.gl/fGg8cyNUFdGQr25b6. Accessed 10 July 2024 at 3:31 PM.

https://images.app.goo.gl/wVxSGDDMKQ7W1wDEA. Accessed 17 July 2024 at 3:31 PM.
https://thiswestcoastmommy.com/6-ways-to-fold-a-prefold-diaper/. Accessed 15 July 2024 at 3: 31 PM.
https://www.allaboutclothdiapers.com/peek-inside-a-pocket-all-in-two-cloth-diaper/. Accessed 17 July 2024 at 3:31 PM.
https://www.charliebanana.com/blog/cloth-diapers-vs-disposable-pros-and-cons/. Accessed 09 Aug 2024 at 4:50 PM.
https://www.charliebanana.com/blog/cloth-diapers-vs-disposable-pros-and-cons/. Accessed 17 July 2024 at 3:31 PM.
https://www.charliebanana.com/blog/types-of-cloth-diapers-cloth-diapering-guide/. Accessed on 22 Aug 2024 at 9:15 AM.
https://www.grandviewresearch.com/industry-analysis/baby-diapers-market. Accessed 22 Aug 2024 at 9:15 AM.
https://www.mordorintelligence.com/industry-reports/baby-diapers-market. Accessed 22 Aug 2024 at 9:15 AM.
https://www.pinterest.com/pin/233342824426333296/. Accessed 20 July 2024 at 3:31 PM.
https://www.simplymombailey.com/a-look-at-the-new-pampers-pure-hybrid/. Accessed 17 July 2024 at 3:31 PM.
https://www.statista.com/outlook/cmo/tissue-hygiene-paper/baby-diapers/worldwide. Accessed 24 Aug 2024 at 9:15 AM.
https://www.statista.com/outlook/cmo/tissue-hygiene-paper/baby-diapers/worldwide#analyst-opinion. Accessed 24 Aug 2024 at 9:15 AM.
https://www.un.org/en/exhibits/page/sdgs-17-goals-transform-world. Accessed 24 Aug 2024 at 9: 15 AM.
https://www.un.org/sustainabledevelopment/news/communications-material/. Accessed 24 Aug 2024 at 9:15 AM.
https://www.undp.org/eswatini/blog/exploring-sustainable-solutions-single-use-diapers. Accessed 24 Aug 2024 at 9:15 AM.
Jordan, W. E., Lawson, K. D., Berg, R. W., et al. (1986). Diaper dermatitis: frequency and severity among a general infant population. *Pediatric Dermatology, 3*, 198–207.
Kabiri, K. (2003). Synthesis of fast-swelling superabsorbent hydrogels: effect of crosslinker type and concentration on porosity and absorption rate. *European Polymer Journal, 39*(7), 1341–1348. https://doi.org/10.1016/S0014-3057(02)00391-9
Khan, T. (2019). A noninvasive smart wearable for diaper moister quantification and notification. *International Journal of Electrical and Computer Engineering (IJECE), 9*(4), 2848–2862. https://doi.org/10.11591/ijece.v9i4.pp2848-2862
Khoo, H. H. (2019). Environmental impacts of conventional disposable diapers versus reusable cloth diapers. *Journal of Cleaner Production, 219*, 655–666.
Khoo, S. C., Phang, X. Y., et al. (2019). Recent technologies for treatment and recycling of used disposable baby diapers. *Process Safety and Environmental Protection, 123*, 116–129. https://doi.org/10.1016/j.psep.2018.12.016
Knowlson, R. (2001). *Airlaid products investments in the future*. http://www.nonwovens-industry.com/articles/2001/10/
Kotz, J. C., Treichel, P. M., & Townsend, J. R. (2009). *Chemistry and chemical reactivity*. Thomson Higher Education.
Kuwayama, M., Kon, Y., & Yamamoto, K. (1992). New etiological factors in diaper dermatitis: Skin irritability and focal irritancy as major causes of diaper dermatitis. *Journal of Paediatric Dermatology, 31*(Suppl 1), 9–14.
Lane, A. T., Rehder, P. A., & Helm, K. (1990). Evaluations of diapers containing absorbent gelling material with conventional disposable diapers in newborn infants. *American Journal of Diseases of Children, 144*, 315–318.
Lörcks, J. (1998). Properties and applications of compostable starch based plastic material. *Polymer Degradation and Stability, 59*(1–3), 245–249. https://doi.org/10.1016/S0141-3910(97)00168-7

Lund, C. H., Kuller, J., Lane, A. T., Lott, J. W., Raines, D. A., & Thomas, K. K. (2001). Neonatal skin care: Evaluation of the AWHONN/NANN research-based practice project on knowledge and skin care practices. *Journal of Obstetric, Gynecologic, & Neonatal Nursing, 30*, 30–40.

Mendoza, J., et al. (2019). Cost analysis of biodegradable diapers. *Sustainability, 11*(20), 5780.

Mishra, L., & Basu, G. (2020). Coconut fibre: Its structure, properties and applications. In *Handbook of natural fibres* (pp. 231–255). Woodhead Publishing.

Mukhopadhyay, S., & Vinay, K. A. (2008). Comparative studies on diaper types. *Environmental Monitoring and Assessment, 144*(1), 223–228.

Newswire. (2021). *Global diaper market analysis*. Newswire.

Ng, C. K., Jap, W., Ong, K. L., & Mok, K. W. (2013). Variety of sizes, shapes, and materials of diapers. *Proceedings of the Institution of Mechanical Engineers, Part E: Journal of Process Mechanical Engineering, 227*(1), 14–22.

Nyamukamba, P., Mququ, Z., Nkosi, S., & Isaacs, S. (2023). Determination of selected harmful substances in baby diapers available on the South African market. *International Journal of Environmental Research and Public Health, 20*(2), 1023. https://doi.org/10.3390/ijerph20021023h

Odio, M., & Friendlander, L. F. (2000). Diaper dermatitis and advances in diaper technology. *Current Opinion in Pediatrics, 12*, 342–346. https://doi.org/10.1097/00008480-200008000-00011

Odio, M., & Thaman, L. (2014). Diapering, diaper technology, and diaper area skin health. *Pediatric Dermatology, 31*(Suppl 1), 9–14.

Odio, M. R., O'Connor, R. J., Sarbaugh, F., & Baldwin, S. (2000). Continuous topical administration of a petrolatum formulation by a novel disposable diaper 1. Effect on skin surface microtopography. *Dermatology, 200*, 232–237.

Odio, M., Jing, W., Baoping, Z., Liu, J., & Guan, Y. (2013). *Fecal coliform contamination in the domestic environment in Chinese infants: Effects of newborn and diaper practice*. Presented at the 2013 excellence in paediatrics conference, Qatar.

Ozen, I., Cincik, E., & Simsek, S. (2016). Thermal comfort properties of simulated multilayered diaper structures in dry and wet conditions. *Journal of Industrial Textiles, 46*(1), 256–278. https://doi.org/10.1177/1528083715580521

Paramanandham, J., Ross, R., Abbiramy, K. S., & Muthulingam, M. (2014). Studies on the moisture retention capacity of coir pith, as a function of time. *International Journal of ChemTech Research, 6*(12), 5049–5052.

Patiño-Masó, J., et al. (2019). Biodegradable materials in diapers. *Waste Management, 87*, 123–130.

Pytlik, E., Molino, D., & Moritz J. (2000). *Superabsorbent polymers*. Available: http://www.eng.buffalo.edu/courses/ce435/Diapers/Diapers.html. Accessed 12 July 2010.

Ramamoorthy, S., et al. (2015). Sustainability and environmental impact of cloth diapers. *Environmental Science and Technology, 49*(3), 1898–1905.

Renukadevi, M., & Poornima, N. (2007). The natural, green and eco-friendly textile materials. *Journal of the Textile Association, 67*(5), 221.

Runeman, B. (2008). Skin interaction with absorbent hygiene products. *Clinics in Dermatology, 26*, 45–51.

Sasikumar, G., Senthil, M., Visagavel, K., Abdul, Z. H., & Dheenathayalan, T. (2014). Development of bio-degradable baby diapers. *International Journal of Research in Engineering and Technology, 3*(Special Issue 11), 186–191.

Shabrin, F. G., & Palaniswamy, N. K. (2019). Development of Diaper using Regenerated Carboxylated Cotton Fiber. *International Journal of Recent Technology and Engineering (IJRTE), 8*(2S11), 2949–2952.

Shanmugasundaram, K., & Gowda, T. (2011). Environmental impact of various diaper types. *Resources, Conservation and Recycling, 55*(9), 840–845.

Shanmugasundaram, O. L., & Gowda Mahendra, R. V. (2010). Development and characterization of bamboo and organic cotton fibre blended baby diapers. *IJFTR, 35*, 201–205.

Šikić, P. M., Maver, U., Marčun, V. N., & Mičetić-Turk, D. (2018). Diagnosis and management of diaper dermatitis in infants with emphasis on skin microbiota in the diaper area. *International Journal of Dermatology, 57*, 265–275. https://doi.org/10.1111/ijd.13748

Srinivas Sahana, M., & Sandipan, D. (2016). Advances in diaper technology. *Indian Journal of Paediatric Dermatology, 17*(2), 83–86.

Stamatas, G. N., & Tierney, N. K. (2014). Diaper dermatitis: etiology, manifestations, prevention, and management. *Pediatric Dermatology, 31*, 1–7. https://doi.org/10.1111/pde.12245

Thilakavathi, P., & Ramakrishnan, G. (2021). Development of organic cloth diapers with disposable coconut coir fibre lining. *International Research Journal of Engineering and Technology, 8*(5), 4701–4705.

Van Gysel, D. (2016). Infections and skin diseases mimicking diaper dermatitis. *International Journal of Dermatology, 55*(s1), 10–13.

Van, R., Morrow, A. L., Reves, R. R., & Pickering, L. K. (1991a). Environmental contamination in child day-care center. *American Journal of Epidemiology, 133*, 460–470.

Van, R., Wun, C. C., Morrow, A. L., & Pickering, L. K. (1991b). The effect of diaper type and overclothing on fecal contamination in day-care centers. *The Journal of the American Medical Association, 265*, 1840–1844.

Ward, D. B., Fleischer, A. B., Feldman, S. R., & Krowchuk, D. P. (2000). Characterization of diaper dermatitis in the United States. *Archives of Pediatrics and Adolescent Medicine, 154*, 943–946.

White, C. F. (2003). Engineered structures for use in disposable incontinence products. *Journal of Engineering in Medicine, 217*, 243–251.

Wiesemann, F., & Adam, R. (2011). Absorbent products for personal health care and hygiene. In V. T. Bartels (Ed.), *Hand book of medical textiles* (Woodhead Publishing series in textiles: Number 100). Wood head Publishing Limited. ISSN 2042–0811 (online).

Yu, J., Treat, J., & Brod, B. (2017). Patch test series for allergic perineal dermatitis in the diapered infant. *Dermatitis, 28*, 70–75. https://doi.org/10.1097/DER.0000000000000256

Curbing Carbon Emissions: Impacts on Environment and Steps to Achieve UN SDGs

Sayantika Debnath and Sankar Roy Maulik

1 Introduction

Climate change caused due to human activities has inflicted detrimental impacts on the overall productivity of agricultural systems, disrupting the delicate balance of ecosystems by forcing various species to migrate from their customary habitats, and is projected to expedite the process of species extinctions on a global scale, as highlighted in the report by the Intergovernmental Panel on Climate Change (IPCC) in 2007. In the realm of political discourse and activity, the issue of climate change has garnered significant attention and recognition, being widely acknowledged as one of the most critical environmental challenges faced by humanity over the last century (Peters et al., 2015).

The textile and apparel sectors stand as pivotal domains within the intricate web of the global economy, encompassing a wide array of activities ranging from the creative design processes to the manufacturing and distribution of various types of fabrics and garments. This industry exerts a profound influence on shaping fashion trends, stimulating economic growth, and generating employment opportunities on a vast scale across different regions of the world. Nevertheless, the textile and apparel industry also bears a considerable environmental footprint, especially regarding its contribution to carbon emissions. The intricate web of textile production involves the substantial consumption of resources such as electricity, fuel, and water, which in turn leads to the release of greenhouse gas emissions and pollutants into the environment. The utilization of energy resources by the textile sector, whether it is in the form of fuel or electricity, plays a crucial role in determining the overall energy consumption patterns within a nation, with a predominant reliance on fossil fuels

S. Debnath · S. Roy Maulik (✉)
Department of Silpa-Sadana, Visva-Bharati (A Central University), Sriniketan, West Bengal, India

S. S. Muthu (ed.), *Sustainable Textile and Apparel Chain Management*, Sustainable Textiles: Production, Processing, Manufacturing & Chemistry,
https://doi.org/10.1007/978-3-031-80240-9_8

further exacerbating the emission of carbon dioxide into the atmosphere. Notably, the production processes within the textile industry heavily rely on electricity for yarn-spinning activities, while the utilization of fuels takes precedence during the textile wet processing stages. Moreover, the widespread adoption of synthetic fibres like polyester significantly adds to the industry's carbon footprint due to the extensive use of petrochemicals in their manufacturing processes. Beyond the production phase, the apparel industry also contributes to environmental degradation through the emissions generated during transportation and logistics operations, as raw materials and finished goods traverse long distances across the globe. The pervasive trend of fast fashion further intensifies this environmental challenge by promoting a culture of rapid production and disposal, thereby increasing waste accumulation and ecological harm. In addition to the substantial energy consumption, the textile manufacturing sector places a significant strain on water resources, particularly during the wet processing stages, resulting in the discharge of contaminated effluent into the environment. This wastewater is often laden with a diverse array of toxic substances, posing a serious threat to the health of rivers and streams if not managed effectively. In some developing nations where textile factories are concentrated, the lack of investment in clean technologies for wastewater treatment has led to the emergence of phenomena like "colourful rivers and streams," symbolizing the extent of pollution caused by the industry. Statistics from reputable sources like the World Bank reveal that a considerable portion of water pollution stemming from industrial activities can be attributed to the dyeing and treatment processes in the textile and apparel sectors, with as many as 72 different toxic chemicals being generated during textile dyeing procedures.

Implementing water conservation measures and adopting strategies to mitigate water pollution should form an integral part of the industry's agenda to enhance the environmental sustainability of its production methods, especially in regions where water resources are scarce. Apart from the significant carbon emissions associated with energy consumption, the combustion of fossil fuels within the textile industry serves as a primary source of air pollutants such as nitrous oxides (NOx), sulphur dioxide (SO_2), and particulate matter (PM) (Hasanbeigi & Price, 2015; Cai & Choi, 2020).

The new sustainable development agenda for 2030 of the United Nations has been in effect since 2016, marking the commencement of 17 sustainable development goals (SDGs). These SDGs effectively outline the fresh aims regarding economic, social, and environmental advancements, encompassing endeavours like eradicating poverty, fostering economic growth, and ensuring environmental preservation, among others. The SDGs necessitate the collective contributions of all global entities, including governments, corporations, non-governmental organizations, and the general public, towards achieving these goals. Reports from the private sector indicate that engaging in sustainable business practices has the potential to generate economic opportunities amounting to as much as US$ 12 trillion. Simultaneously, there is evidence of prominent fashion corporations proactively aligning themselves with the United Nations' SDGs. Inditex, a Spanish clothing brand meticulously coordinates its sustainable supply chain management following the United

Nations' SDGs, striving diligently to implement strategies for each specific goal. One notable practice of Inditex involves rigorously enforcing the Code of Conduct and evaluating compliance through audits to safeguard the health and well-being of workers. Similarly, H&M emphasizes that its operations must actively contribute to the sustainable development goals, as evidenced by its comprehensive 'Agenda 2030' and aims to exclusively utilize recycled or sustainably sourced materials in its products to ensure environmentally responsible production processes. Furthermore, the prevalence of fast fashion and rapidly changing trends in the fashion industry has led to frequent consumer purchases, shortened apparel lifespans, and a substantial amount of post-consumer textile waste. The textile and apparel sectors are facing a significant challenge in achieving new economic, social, and environmental objectives. Presently, there are emerging concerns surrounding the sustainability implications of fast fashion practices. The rapid turnover of fashion items has resulted in increased waste generation and heightened environmental strain. In response to external pressures, fashion brands have implemented sustainable measures (Dissanayake & Sinha, 2015; Biswas et al., 2018).

In recent years, sustainable practices have increasingly become a focal point within the fashion industry, driven by a growing awareness of environmental and ethical issues. Key players like H&M, Levi's, Gucci, Stella McCartney, and others have implemented various initiatives to address these concerns across different aspects of their operations. H&M, for example, has made significant strides in sustainable cotton sourcing, aiming for 100% sustainable and recycled materials across all its brands by 2030. This commitment includes initiatives to promote organic cotton, which reduces the environmental impact compared to conventional cotton farming methods. By prioritizing sustainable sourcing, H&M seeks to minimize the ecological footprint associated with its cotton consumption. Levi's has also taken notable steps to reduce water consumption and pollution in denim production through innovative technologies. Their 'Waterless Dyeing Process', introduced in 2016, significantly decreases the amount of water required for dyeing denim fabric, thereby mitigating water pollution. Additionally, Levi's Eureka Lab has developed advanced laser technology for distressing and washing jeans, further enhancing eco-efficiency by reducing chemical usage and energy consumption in these processes. In terms of ethical considerations, luxury brand Gucci announced a cessation of genuine leather and fur usage in 2017, reflecting a commitment to animal welfare. This decision underscores a broader industry trend towards cruelty-free alternatives and sustainable materials in fashion. Similarly, Stella McCartney has positioned itself as a pioneer in sustainable fashion with its exclusive use of artificial leather, which offers a high-quality and functional alternative to traditional leather without the environmental and ethical drawbacks. Beyond materials, some companies like ECOALF have innovatively addressed plastic pollution by repurposing materials such as PET bottles into fashionable clothing and footwear. This approach not only reduces waste but also raises awareness about ocean pollution and the importance of recycling in fashion production.

Closing the loop on fashion consumption has become a priority for many brands. H&M's initiative to collect used apparel for recycling, Uniqlo's post-consumer product donation programme, and Nike's recycling of old shoes into new ones demonstrate efforts to minimize textile waste and extend the life cycle of products. These examples illustrate a broader shift towards sustainability in fashion, where companies are increasingly integrating environmental and ethical considerations into their business models. By adopting these practices, brands not only respond to consumer demand for eco-friendly products but also contribute to global efforts to mitigate environmental impact and promote responsible consumption. As sustainability continues to gain prominence in the fashion industry, these initiatives serve as benchmarks for future innovation and collaboration towards a more sustainable future (Cai & Choi, 2020).

Emerging technologies capable of enhancing the efficiency of energy use within the textile industry have the potential to yield various co-benefits, such as decreased air pollution and enhanced human health. Quantifying these additional advantages in monetary terms can enhance the cost-effectiveness of these emerging technologies, particularly considering their higher initial capital costs during the adoption phase.

The global population is projected to increase by approximately 35% by the year 2050 according to the United Nations Department of Economic and Social Affairs (World Population Prospects—Population Division—United Nations, n.d.). This population growth, coupled with the continual economic expansion in developing nations, will lead to substantial rises in both textile production and consumption. Consequently, there will be significant escalations in absolute energy consumption, water usage, and emissions of carbon dioxide (CO_2) and other environmentally detrimental substances by the textile industry.

Numerous research works have examined the existing energy-efficient technologies available for commercial use in the textile sector (Hasanbeigi, 2010). However, meeting future targets for reducing energy consumption and CO_2 emissions in the textile industry will necessitate innovative solutions that surpass the currently available technologies. While some studies have pinpointed energy-efficient technologies already in use within the textile sector, other sources like the US Department of Energy's Advanced Manufacturing Office (n.d.) have highlighted general efficiency technologies that could be adapted for textile production. Nonetheless, there is a scarcity of comprehensive and easily accessible information on alternative, emerging, or advanced energy-efficient, low-carbon, water-conservation, and pollution-reduction technologies that have not yet reached full commercialization within the textile industry.

Efforts to mitigate carbon emissions stemming from the textile and apparel industries demand a holistic strategy that encompasses sustainable practices, enhanced energy efficiency, and the promotion of renewable energy sources. Progress in material innovations, such as organic cotton and recycled fibres, alongside advancements in manufacturing processes, offer potential avenues for diminishing the industry's carbon footprint. Collaboration among various stakeholders, ranging from manufacturers to consumers, plays a pivotal role in propelling substantial transformations and ensuring a more sustainable trajectory for the textile and apparel sector.

This chapter attempts to offer an extensive examination of the substantial contributions of the textile and apparel sectors to carbon emissions and their ecological repercussions. It starts by identifying the problem within the framework of climate change and the United Nations Sustainable Development Goals (SDGs), with particular focus on SDG 13 (Climate Action), SDG 12 (Responsible Consumption and Production), and SDG 8 (Decent Work and Economic Growth). The analysis investigates various phases of the textile manufacturing process—from the cultivation of raw materials, such as cotton to energy-intensive industrial operations, including spinning, weaving, dyeing, synthetic fibre production, and logistics—all of which significantly contribute to the industry's considerable carbon footprint. This chapter also emphasizes the sector's substantial dependence on fossil fuels and its contribution to pollutant generation, especially during wet processing phases and through the distribution of products. The challenges of waste and pollution, encompassing water contamination and the insufficient management of textile effluent in specific locales, are also examined, along with the influence of fast fashion in worsening environmental degradation, prompting increased consumption, diminishing product lifespans, and producing considerable volumes of waste. Attention is brought to the progressive initiatives undertaken by prominent industry figures such as H&M, Levi's, Gucci, and Stella McCartney, which are adopting sustainable practices and innovations to mitigate environmental effects. These initiatives exemplify the utilization of recycled materials, minimizing water and chemical consumption, advocating for renewable energy, and incorporating circularity within production processes. Ultimately, the chapter calls for a holistic approach towards sustainability in the fashion and textile sectors, highlighting the necessity of inter-sectoral collaboration, investment in clean technologies, and embracing innovative approaches to diminish carbon emissions and promote environmental accountability.

2 The Textile Industry and Its Effects on Climate Change

The textile industry is a major contributor to climate change, significantly impacting the environment through its production processes and waste generation as shown in Fig. 1. From the cultivation of raw materials like cotton, which requires huge amounts of water and pesticides, to the energy-intensive manufacturing processes, the industry's carbon footprint is substantial. The dyeing and finishing stages release harmful chemicals into waterways, contributing to water pollution. Moreover, the disposal of textile waste, often through incineration or landfill, releases greenhouse gases (GHGs) and toxins. As a result, the textile industry plays a pivotal role in global greenhouse gas emissions, necessitating urgent action towards sustainable practices and innovations to mitigate its environmental impact.

Climate change arises from the escalation of greenhouse gas emissions, subsequent global warming, and the reaction of the climate system to this warming. The textile industry, due to its intricate supply chain concerning the transportation of

Fig. 1 Pollution through waste generation. (Source: Freepik)

fabrics and garments from manufacturing hubs, typically located in developing nations as well as its energy-intensive production processes, is responsible for emitting approximately 1.2 billion tonnes of CO_2 equivalent, which accounts for nearly 10% of global greenhouse gas emissions (Masson-Delmotte et al., 2021; United Nations, 2018).

The continuous demand for apparel, exacerbated by the prevalence of the fast fashion business model, necessitates a steady flow of natural resources. Projections indicate that by 2050, the fashion sector will contribute to roughly 25% of the world's carbon budget. The increased consumption of natural resources to meet the fast fashion industry's expansion will lead to heightened ecological harm and the exacerbation of climate change. Global fibre production nearly doubled between 2010 and 2020, escalating from 58 million to 109 million tonnes (Berg & Karl-Hendrik, 2021; Leal Filho et al., 2022).

A multi-regional ecologically extended input-output model focusing on the environmental impact of fast fashion and apparel consumption shows that there has been a 23% increase in greenhouse gas emissions from 2000 to 2015. The shorter lifespan of garments before disposal in landfills or through incineration has led to a rise in carbon emissions stemming from both manufacturing processes and the production of materials utilized by the fashion industry. The utilization of synthetic fibres has surged, with polyester now surpassing cotton as the most prevalent fabric in fashion (Niinimäki et al., 2020). However, carbon emissions from synthetic fibres exceed those of cotton due to their derivation from fossil fuels like unprocessed petroleum products, thereby contributing to climate change through carbon emissions. It is estimated that producing one polyester t-shirt emits 5.5 kg of CO_2, compared to 2.1 kg of CO_2 for a cotton t-shirt (Kirchain et al., 2015).

The impetus to increase fabric production to meet the growing manufacturing demands of the fashion industry also plays a role in the acceleration of deforestation. Rayon, a semi-synthetic fibre produced from refined cellulose, alongside similar fabrics like viscose and modal, is sourced from wood pulp. The rayon/viscose sector is anticipated to grow in response to escalating demand, as projected by Mordor Intelligence (Viscose Staple Fibre—Market & Industry Manufacturers, 2024), potentially heightening the threat to forests on a global scale. Approximately 30% of rayon and viscose utilized by the fashion sector are obtained from ancient and endangered forests (McCullough, 2014). Deforestation, the second leading cause of greenhouse gas emissions worldwide, has intensified the occurrence of extreme heat in moderate zones of North America and Europe, according to Stoy in 2018. A similar impact might also be observed in developing regions in the southern hemisphere, though its specific contribution to extreme weather severity remains scientifically uncertain. The rapid turnover of fashion trends and shortened product life cycles result in a substantial growth of post-consumer textile waste, significantly adding to landfill expansion and incineration, leading to the release of harmful gases. Absent a comprehensive recovery and recycling strategy, the environmental repercussions are likely to worsen, especially considering the substantial amount of resources, water, and energy invested in the production of new textile goods, the majority of which ultimately end up as textile waste once discarded by consumers.

It is evidently apparent that the textile sector possesses a significant carbon footprint, thus making a noteworthy contribution to the overall global greenhouse gas emissions. Each stage of the textile supply chain, starting from the cultivation of raw materials to the processes of manufacturing, dyeing, and transportation, is characterized by a high level of energy consumption. The cultivation of cotton, which serves as a pivotal raw material in this industry, heavily depends on both water resources and the use of pesticides, whereas synthetic fibres such as polyester are predominantly sourced from petroleum, which is a non-renewable fossil fuel. The processes of dyeing and finishing often entail the utilization of hazardous chemicals and require substantial quantities of water and energy, thereby exacerbating the environmental impact. Furthermore, the transportation of textile products across extensive global supply chains contributes significantly to carbon emissions. The concept of fast fashion, known for its rapid production cycles and affordable clothing items, further intensifies these adverse effects, leading to increased levels of waste generation and pollution. The overall carbon footprint of the textile industry emphasizes the urgent need for the implementation of sustainable practices and the development of innovative solutions to mitigate the environmental repercussions associated with its operations.

3 Carbon Footprint

3.1 Definition

The carbon footprint concept originates in the broader ecological footprint theory, which William Rees and other scholars have introduced. The ecological footprint represents the biologically productive land and sea area required to continuously provide the resources a person, population, or activity consumes and to absorb the waste they generate with the prevailing technology and resource management practices. This measure helps to quantify the demand placed on the Earth's ecosystems and compare it to the planet's capacity to regenerate those resources and absorb wastes, thereby highlighting the sustainability of human activities. Within this framework, the carbon footprint specifically focuses on the amount of land area required to absorb the carbon dioxide emissions produced by human activities, particularly the burning of fossil fuels. By isolating carbon emissions, the carbon footprint provides a clear and tangible measure of one of the most significant contributors to global climate change. This concept has become a crucial tool for understanding and managing the impact of various activities, from individual behaviours to industrial processes, on the Earth's climate system. It allows for the assessment of the carbon intensity of different products, services, and lifestyles, thereby enabling individuals, businesses, and policymakers to make more informed decisions aimed at reducing carbon emissions and mitigating their environmental impact (Finkbeiner, 2009; Zhang et al., 2020).

Carbon footprints typically encompass the direct and indirect emissions of the six greenhouse gases resulting from the production and activities within a specific timeframe, often expressed in terms of CO_2 equivalent. Carbon consumption is substantial, and CO_2 emissions are on the rise.

The discourse surrounding carbon footprints is shaped by three predominant perspectives, each offering a nuanced understanding of what constitutes a carbon footprint. The first viewpoint focuses narrowly on the carbon dioxide (CO_2) emissions produced directly from the combustion of fossil fuels in human activities. This perspective is concerned with the immediate and tangible output of CO_2 resulting from activities such as driving cars, operating machinery, and heating buildings. It emphasizes the direct impact of fossil fuel consumption on the atmospheric concentration of CO_2, which is the primary greenhouse gas driving global warming.

The second perspective broadens the definition of a carbon footprint to include not only CO2 emissions but also other greenhouse gases (GHGs) such as methane (CH_4), nitrous oxide (N_2O), and fluorinated gases. These emissions are converted into CO_2 equivalents (CO_2e) to provide a standardized measure of their global warming potential. This viewpoint considers the entire life cycle of products, from raw material acquisition and production to distribution, utilization, and end-of-life recovery. By encompassing all stages of a product's life cycle, this perspective captures the total GHG emissions associated with a product, providing a comprehensive assessment of its environmental impact.

The third perspective emphasizes the importance of evaluating the impact of human activities on climate change by considering both direct and indirect CO_2 conversions. This holistic approach extends the assessment beyond immediate emissions to include the broader implications of various entities' actions, such as individuals, populations, governments, businesses, and industry sectors. It recognizes that carbon footprints are not limited to direct (on-site, internal) emissions but also encompass indirect (off-site, external, embodied, upstream, downstream) emissions. For example, indirect emissions might arise from the production of goods and services used by an organization, the transportation of raw materials, or the disposal of products at the end of their life cycle. By accounting for all forms of emissions, this perspective provides a more accurate and encompassing measure of the true impact of human activities on climate change.

The concept of a carbon footprint serves as a fundamental measure in assessing the environmental impact of human activities, particularly through the emission of carbon dioxide (CO_2). It provides a clear starting point for understanding the contribution of various actions, products, or processes to global warming and climate change. However, this metric has limitations when considering other greenhouse gases (GHGs) like methane, which also significantly contribute to climate change albeit through different mechanisms and with varying potency. The challenge lies in the complex task of quantifying these gases accurately due to data limitations or their non-carbon basis, which complicates their aggregation into a single metric alongside CO_2. While including methane could provide a more comprehensive view, creating a universally accepted aggregate metric poses practical difficulties and may dilute the clarity of the carbon footprint concept. Advocates for a broader metric propose a 'climate footprint' that encompasses all relevant GHGs, providing a more holistic measure of environmental impact. Yet, for simplicity and practicality, the focus remains primarily on CO_2 emissions in the widely used carbon footprint calculation (Wiedmann & Minx, 2008).

Recently, the European Commission made a significant announcement mandating Member States to decrease their greenhouse gas emissions by a minimum of 20% before 2020 in comparison to 1990 levels. This reduction target was predicted to potentially escalate beyond 20% if other industrialized nations commit to similar endeavours. This directive holds implications for not only the European Union Member States but also for various stakeholders, including globally operating enterprises that have relocated a substantial portion of their production operations to low-cost countries and have established intricate intermodal transportation networks to cater to their markets (Halldórsson et al., 2009).

An initial stride taken towards curbing greenhouse gas emissions involves the introduction of the carbon footprint as a metric, defined as 'the total amount of carbon dioxide and other greenhouse gases emitted over the entire life cycle of a product or service' (Gereffi, 2001). The standard unit for measuring carbon footprint typically revolves around tons of CO_2 and serves as a tool to gauge the environmental impact of a product or service in relative terms (Bevilacqua et al., 2012).

3.2 *Types of Carbon Footprints*

Two distinct categories of carbon footprint are commonly recognized in environmental assessments:

1. Organizational carbon footprint
2. Product/service carbon footprint

The organizational carbon footprint evaluation involves the comprehensive measurement of greenhouse gas (GHG) emissions stemming from all operational activities within a company. This encompasses the energy consumption in buildings, industrial processes, company-owned vehicles, and possibly the assessment of indirect emissions linked to activities beyond the organization's immediate operations, known as the value chain. Value chain analysis is a methodical examination of every phase a business undergoes, starting from the procurement of raw materials to the final delivery to end-users.

On the other hand, a product's carbon footprint pertains to the calculation of GHG emissions throughout the entire lifespan of a specific product or service. This intricate process includes the assessment of emissions originating from the extraction of raw materials, and manufacturing procedures, as well as the emissions associated with the usage and disposal of the product. The implementation of carbon footprints aids in recognizing and quantifying the primary sources of emissions, facilitating the identification of opportunities to mitigate carbon outputs within an organization. This strategic approach enables entities to monitor and regulate carbon emissions by establishing reduction targets and tracking progress, which is tailored to suit the unique requirements of the business.

Adhering to an environmental standard is imperative to ensure the utilization of a consistent and reliable methodology when computing carbon emissions. It is essential to select the most pertinent and effective approach for an organization based on the available data and the intended application of the outcomes. The choice of standard depends on several factors, including the level of data accessibility and the specific purpose of the results obtained (Peters et al., 2015).

3.2.1 Organizational Footprinting Standards

3.2.1.1 ISO 14064-1

The ISO 14064 standard, published in 2006, is an integral part of the ISO series of International Standards for environmental management. This series of standards is designed to help organizations manage their environmental responsibilities systematically and effectively. The ISO 14064 standard specifically focuses on the reporting of greenhouse gas (GHG) emissions, offering comprehensive guidelines and principles to ensure accurate and transparent reporting. ISO 14064 provides detailed guidance on the principles and requirements for reporting GHG emissions, ensuring that organizations can consistently and accurately measure and report their emissions. This is crucial for maintaining transparency and accountability in environmental reporting. The standard covers all aspects of GHG emissions reporting, from the initial measurement and data collection to the final reporting and verification. In addition to the core principles and requirements, ISO 14064 offers further guidance on the verification process. This includes the necessary levels of data validation to ensure that the reported emissions are accurate and reliable. The standard also outlines external reporting frameworks that organizations can use to communicate their emissions data consistently to external stakeholders. This aspect of the standard is particularly important for organizations that need to report their emissions data to regulatory bodies, investors, or the public. By providing a standardized approach to GHG emissions reporting, ISO 14064 helps organizations to ensure that their emissions data is comparable and consistent with that of other organizations. This facilitates better benchmarking and performance tracking, allowing organizations to identify areas for improvement and to demonstrate their commitment to reducing their environmental impact. Overall, ISO 14064 plays a critical role in promoting transparency and accountability in GHG emissions reporting, contributing to global efforts to manage and reduce greenhouse gas emissions.

3.2.1.2 GHG Protocol Corporate Standard

The GHG Protocol is a widely used emissions accounting tool employed by numerous businesses and organizations around the world. Developed to meet the need for a consistent approach in corporate carbon accounting and reporting, the GHG Protocol provides a comprehensive framework for measuring and managing greenhouse gas emissions. This standard is essential for ensuring that organizations can accurately track their emissions and implement effective strategies to reduce their carbon footprint. A key component of the GHG Protocol is the Corporate Standard, which categorizes GHG emissions into three distinct scopes. These scopes provide a clear structure for understanding and reporting the different types of emissions that an organization may generate.

Scope 1 (Direct emissions that result from activities within your organisation's control)

Scope 2 (Indirect emissions from any electricity, heat, or stream you purchase and use)

Scope 3 (other indirect emissions from sources outside you direct control)

3.2.1.3 GHG Protocol Value Chain Standard

The standard accompanying the GHG Protocol Corporate Standard offers additional guidance for companies aiming to assess their entire value chain Scope 3 emissions.

While the Corporate Standard provides a foundation for measuring direct (Scope 1) and indirect (Scope 2) emissions, this complementary standard focuses on the often complex and extensive Scope 3 emissions. These emissions occur as a result of activities both upstream and downstream of a company's operations, making their assessment critical for a comprehensive understanding of an organization's overall environmental impact.

Scope 3 emissions are grouped into 15 categories, encompassing a wide range of activities that contribute to a company's carbon footprint. These categories include upstream activities such as purchased goods and services, capital goods, fuel- and energy-related activities, and transportation and distribution. They also cover downstream activities like the processing of sold products, the use of sold products, the end-of-life treatment of sold products, and downstream leased assets. By categorizing Scope 3 emissions in this manner, the standard helps companies systematically identify and measure all relevant indirect emissions sources within their value chain.

This detailed approach to Scope 3 emissions is particularly useful for companies that wish to report comprehensive information on their value chain emissions externally. Organizations reporting to the Carbon Disclosure Project (CDP), for instance, are required to provide extensive assessments of their value chain emissions. The guidance provided by this standard facilitates accurate and thorough reporting, ensuring that companies can meet such requirements and demonstrate their commitment to sustainability.

3.2.2 Product Footprinting Standards

3.2.2.1 PAS 2050

PAS 2050, developed by the British Standards Institute (BSI), emerged from the need for a consistent method to assess the life cycle greenhouse gas (GHG) emissions associated with products and services. Recognizing the growing concern about climate change and the impact of carbon emissions, BSI aimed to provide a standardized approach that businesses and organizations could use to measure and manage the carbon footprint of their offerings. This standardization helps ensure that assessments are reliable and comparable across different products and services, facilitating more informed decisions and actions towards sustainability.

The standard has gained wide recognition and international application, reflecting its robust methodology and the growing global emphasis on reducing GHG emissions. By providing a consistent framework, PAS 2050 allows for the comprehensive evaluation of product life cycle GHG emissions. This consistency is crucial as it enables various stakeholders, including consumers, businesses, and policymakers, to understand and compare the environmental impact of different products and services accurately. The international acceptance of PAS 2050 underscores its credibility and utility in global efforts to address climate change.

PAS 2050 is versatile and can be applied to a broad spectrum of product and service types. Whether it is goods or services, business-to-consumer transactions, or business-to-business operations, the standard offers a valuable tool for manufacturers, retailers, and traders. This wide applicability ensures that diverse industries can adopt the standard to measure and manage their carbon footprints effectively, driving broader environmental benefits across different sectors of the economy.

One aspect of PAS 2050 is that it does not prescribe specific requirements for external communication of a product's carbon footprint. While it provides the methodology for assessing GHG emissions, it leaves the decision of how to communicate this information to external stakeholders at the discretion of the organization. This flexibility allows businesses to tailor their communication strategies to their specific contexts, regulatory environments, and stakeholder needs, while still benefiting from the rigorous assessment framework provided by PAS 2050.

3.2.2.2 GHG Protocol Product Lifecycle Accounting and Reporting Standard

The GHG Protocol Product Lifecycle Accounting and Reporting Standard was launched in October 2011, following an extensive three-year development process that involved multiple stakeholders. This collaborative approach ensured that the standard incorporated a wide range of perspectives and expertise, making it comprehensive and practical for various industries. The development process included input from businesses, non-governmental organizations, government agencies, and academic institutions, reflecting a concerted effort to address the complexities of measuring and reporting greenhouse gas (GHG) emissions throughout a product's life cycle.

This standard builds upon existing ISO environmental guidance, integrating well-established principles and practices into its framework. By leveraging the foundational work of ISO standards, the GHG Protocol standard aims to provide a consistent and reliable method for accounting and reporting product life cycle GHG emissions. This integration ensures that the standard is grounded in internationally recognized environmental management practices, enhancing its credibility and acceptance worldwide.

The primary objective of the GHG Protocol Product Lifecycle Accounting and Reporting Standard is to offer a general framework for businesses and organizations to systematically account for and report the GHG emissions associated with their products. This framework encompasses the entire life cycle of a product, from raw material extraction through production, distribution, use, and end-of-life disposal. By considering emissions at each stage of the product life cycle, the standard enables a comprehensive assessment of a product's overall environmental impact.

A key requirement of the standard is public reporting. To claim conformance to the GHG Protocol Product Lifecycle Accounting and Reporting Standard, organizations must publicly disclose their GHG emissions data. This transparency is crucial for several reasons. It ensures accountability, as stakeholders can verify that the organization is accurately measuring and reporting its emissions. Public reporting also fosters trust and credibility with consumers, investors, and other stakeholders who are increasingly concerned about the environmental impact of products and services. By making this information publicly available, organizations demonstrate their commitment to sustainability and provide valuable data that can inform decision-making and drive further environmental improvements.

3.2.2.3 ISO 14067

The primary purpose of ISO 14067 is to enhance transparency in the reporting of greenhouse gas (GHG) emissions associated with the entire life cycle of various products and services. This standard addresses the growing demand from consumers, businesses, and regulatory bodies for clear and accurate information on the environmental impact of products. By promoting transparency, ISO 14067 aims to ensure that GHG emissions data is readily accessible, allowing stakeholders to make informed decisions and encouraging organizations to adopt more sustainable practices.

ISO 14067 is closely aligned with previous ISO standards and PAS 2050, which are also focused on product carbon footprinting. This alignment is intentional, as it seeks to increase the comparability of product carbon footprints on an international scale. By harmonizing the methodologies and guidelines used in different standards, ISO 14067 ensures that GHG emissions data is consistent and comparable across various products and regions. This comparability is crucial for global efforts to reduce GHG emissions, as it enables businesses and policymakers to benchmark performance, set targets, and track progress in a standardized manner. In addition to its focus on transparency and comparability, ISO 14067 provides detailed guidance on the requirements for public reporting and external communication of product carbon footprints. This includes specifying the information that needs to be disclosed, the formats for reporting, and the communication channels that can be used. By offering clear instructions on public reporting, the standard helps organizations effectively communicate their GHG emissions data to stakeholders, including consumers, investors, and regulatory authorities. This transparency not only builds trust but also supports accountability and drives continuous improvement in environmental performance.

ISO 14067 includes additional guidance on the verification and assurance of product carbon footprints. Verification and assurance processes are essential for ensuring the accuracy and credibility of GHG emissions data. By providing detailed instructions on how to conduct these processes, the standard helps organizations validate their carbon footprint calculations and gain third-party assurance. This independent verification enhances the reliability of the reported data, providing stakeholders with confidence that the information is accurate and trustworthy.

Through an examination of the existing body of literature focusing on the life cycle of the textile industry, it becomes evident that there is a scarcity of scholarly attention dedicated to this particular area. Only a limited number of researchers have delved into studying the life cycle aspects of textile clothing and regional textile production processes. This deficiency in research is particularly noticeable when considering the evaluation of carbon footprints within the textile industry of India. Data reveals that the energy consumption throughout the entirety of the textile industry's production procedures, encompassing the life cycle from fibre processing to the creation of clothing products, amounts to approximately 4.84 tons of standard coal per ton of fibre. This value represents roughly 4.4% of the total energy consumption across all industrial sectors within the nation (Zhang et al., 2020). In light of the escalating concerns surrounding global warming, the textile industry must address the unsustainable patterns of high energy consumption and significant pollution levels. Urgent measures are required to curtail the proliferation of outdated production practices, reconfigure industrial frameworks, and transition towards sustainable developmental pathways. Consequently, conducting research focused on evaluating the carbon footprint of the textile industry through a comprehensive life cycle perspective holds considerable significance in fostering advancements in energy efficiency and emission reduction initiatives.

4 Distribution of Emissions by Geographical Region

Limited data exists regarding the regional distribution of emissions stemming from the textile and garment sectors. Nevertheless, it is feasible to deduce the locations where substantial emissions are generated by examining the pivotal hubs for the production and consumption of textiles and garments.

4.1 Production-Related Emissions

When considering emissions from production, it is noteworthy that China stands as the foremost global producer of textiles and garments, with the apparel industry ranking as the sixth-largest sector concerning energy consumption (Zhang & Chen, 2019). In Bangladesh, the textile and garments industry serve as the primary export earner, contributing to 81% of the country's gross domestic income. The industrial domain accounts for 27% of the nation's overall energy usage, and within this sector, the textile industry comprises approximately one-third of its consumption (Hasan et al., 2019). Meanwhile, in Turkey, another prominent textile-producing nation, the sector emerges as the third-most energy-intensive realm following iron/steel and cement (Alkaya & Demirer, 2014). The energy origins in production countries significantly influence the emission intensity associated with textiles and garment manufacturing. For instance, Bangladesh heavily relies on natural gas, with renewable energy sources making up less than 2% of the total energy mix (Hasan et al., 2019).

Conversely, within nations devoid of large garment production hubs, the emissions deriving from the textile and garment sector remain considerably low. For example, in the United Kingdom of Great Britain and Northern Ireland, the textile and garment industries are estimated to contribute to less than 1% of the total emissions (Allwood et al., 2015).

4.2 Consumption-Related Emissions

Shifting the focus to emissions arising from consumption, aside from emissions associated with transporting goods to the final consumer, which are deemed insignificant, greenhouse gas emissions are generally attributed to the country where the goods or services are manufactured. Nonetheless, there exists contentions advocating for the allocation of emissions to the countries engaging in the consumption of said goods or services (Peters & Hertwich, 2008). Adhering to this perspective, it is approximated that per capita emissions stemming from clothing consumption amount to 1450 kg CO_2 annually in the United States, 1210 kg CO_2 in Europe, and 41.8 kg CO_2 in China (Quantis, 2018). A research endeavour conducted in Sweden approximated emissions originating from clothing consumption to be 330 kg CO_2 per individual per year, equating to roughly 3% of the individual's carbon footprint (Sandin et al., 2019).

5 Emissions Distribution Along the Textile Value Chain

Quantifying the distribution of carbon emissions across the value chain poses a significant challenge due to its reliance on the specific product, materials involved, and the emissions intensity of the country where production occurs (WRI 2019). The complexity of this task is further compounded by the multitude of variables at play. It is crucial to acknowledge the diverse methodologies, scopes, and geographic locations of these studies, resulting in a lack of consensus among their outcomes. Nonetheless, a common thread running through most studies is the identification of yarn and textile production, along with the use phase, as the primary contributors to emissions. Conversely, the production of raw materials, assembly, distribution, and disposal are associated with comparatively lower emission shares.

5.1 Production of Clothing and Textiles

A singular study stands out for its comprehensive assessment of emissions within the global garment sector. In 2018, Quantis undertook this ambitious task and revealed that the dyeing and finishing processes accounted for the largest proportion of emissions at 36%, with yarn preparation following closely at 28%. Furthermore, fibre production contributed 15%, while fabric preparation constituted 12% of total emissions. Surprisingly, assembly activities were responsible for a mere 7% of emissions, with distribution and end-of-life disposal registering negligible amounts of emissions according to the study. Omissions in the analysis of the use phase and customer transport were attributed to the inherent challenges in capturing the variability in consumer behaviour (Quantis, 2018). These findings align with a Swedish study examining emissions from six garment types, where wet treatment processes, specifically dyeing and finishing, emerged as the predominant emission source, accounting for 23.5% of emissions (Sandin et al., 2019).

The Swedish study assumed the utilization of the best available technologies in textile manufacturing, a scenario that diverges from the reality of industry practices, potentially resulting in unaccounted emissions during garment production. Studies focusing on specific factories, brands, or garment types echoed similar trends. Notably, a 2019 investigation of a men's shirt sewing assembly line in China identified fabrics as the primary carbon emission source (Zhang & Chen, 2019). Similarly, a study on the carbon footprint of cotton T-shirt production in China highlighted the dyeing process as the largest contributor to emissions at 35%, with garment assembly also playing a significant role at 32% (Zhang et al., 2015). H&M's 2019 study underscored fabric production, inclusive of yarn production and dyeing/finishing, as the most substantial emissions source at 48%, followed by the use phase at 13%, garment assembly at 12%, and raw materials at 11%. These diverse studies collectively contribute to a nuanced understanding of carbon emissions distribution across the garment industry landscape.

The substantial proportion of emissions incurred during these stages can be directly attributed to the considerable energy demand of these processes and the heavy reliance on fossil fuels for energy in the countries where production takes place. Wet textile processing, a method frequently employed for dyeing purposes, stands out as a highly energy-intensive procedure which typically necessitates the direct utilization of coal or natural gas on-site for steam generation. For instance, a comprehensive examination of a cotton mill in Turkey revealed that the primary energy consumption arose from natural gas employed for steam generation, accounting for 46% of the total energy usage, predominantly for heating water in the wet processing phase. Additionally, the second-largest energy consumption was allocated to drying activities, constituting 30% of the energy consumption at the mill (Alkaya & Demirer, 2014). Another study conducted in Bangladesh highlighted substantial energy consumption in the wet processing unit, where a combination of steam and heat energy is utilized, and in the weaving machines, particularly looms, which consume between 50% and 60% of the energy at the weaving facility (Hasan et al., 2019). Emissions during the dyeing phase of cotton T-shirt production in China were primarily associated with the on-site combustion of coal for steam generation, while emissions during the garment assembly phase were indirect, stemming from electricity consumption within the factory premises (Zhang et al., 2015).

5.2 Consumer Use Phase

Analysing the contribution of the utility phase by textile and garment consumers to the carbon emissions throughout the life cycle of a garment is not a common practice. The calculation of emissions from the use phase presents challenges due to the scarcity of data concerning consumer behaviours such as washing frequency, water temperature, choice of detergents, and drying methods, which exhibit significant variations across different cultures (Yasin et al., 2016). Some research studies propose that the use phase could potentially be the most significant contributor to emissions within the value chain. A study commissioned by the UK brand Marks & Spencer (M&S) identified the use phase as the most energy-intensive stage throughout the life cycle, accounting for 81% of energy consumption for men's cotton briefs and 76% for polyester trousers. Notably, washing and drying activities emerged as the primary energy consumers and hence major sources of emissions. In both cases, manufacturing processes accounted for 13%, retail operations for 4%, while cotton fibre production only contributed 3% compared to 7% for polyester (Environmental Resources Management 2002). Levi Strauss & Co also observed that the use phase was responsible for the largest share of emissions at 34%, surpassing fabric production (31%) and cotton cultivation (10%). An investigation in China similarly uncovered substantial emissions during the use phase, ranking it as the third-highest contributor to emissions. Nonetheless, this study was based on washing practices in China, which are less energy-intensive compared to other major economies, implying that emissions could be higher in different countries (Zhang et al., 2015).

5.3 Distribution and End-of-Life

The various scholarly inquiries aforementioned have all come to the consensus that the sector of transport does not play a significant role as an energy consumer within the overall value chain, with estimations ranging from 1% to 6%. Nevertheless, an examination conducted by M&S pointed out that the transportation undertaken by a consumer when travelling from the store consumes more energy than all other forms of transport involved in the value chain, a crucial aspect that was not taken into consideration in the Quantis study. Furthermore, the investigation carried out in Sweden regarding six different garments revealed a noteworthy portion of emissions stemming from transport during the utilization phase, constituting slightly over 10% of the total emissions. The disposal phase was also scrutinized, indicating a negligible impact on emissions, with estimations varying between 0% and 2.8% across the array of studies scrutinized. Nonetheless, the approach of life cycle assessment (LCA) has been a focal point in several research endeavours aimed at assessing the environmental repercussions of clothing sharing, reuse, and recycling practices. For instance, a study conducted in Finland unveiled that augmenting the collection of discarded textiles for reuse and recycling could potentially yield environmental advantages by counterbalancing virgin textile production. Nevertheless, there exists uncertainties regarding the capacity of recycled fibres to substitute virgin fibres, as well as the effects of textile reuse and resale on the demand for new production or its impact on overall consumption levels (Dahlbo et al., 2017). The investigation by M&S exemplified the environmental advantages associated with clothing reuse, elucidating that the processing and distribution of second-hand clothing consume approximately 1.7-kilowatt hours (kWh) of energy per kilogram, and each kilogram of clothing substituted by second-hand items could potentially save 65 kWh for cotton and 90 kWh for polyester, as reported by Environmental Resources Management in 2002. A study focusing on clothing libraries in Sweden indicated that under certain scenarios, the emissions generated from transportation may overshadow the environmental benefits arising from the reduced production of new clothing. The pivotal factors that safeguard an environmental advantage in such contexts include prolonging the lifespan of garments and opting for transportation methods with minimal environmental impact, as underscored in the research by Zamani, Sandin, and Peters (2017).

5.4 Variations Between Fibres

The impact of the type of fibre on greenhouse gas (GHG) emissions is a crucial consideration within the textile industry, although the availability of published data on this topic is currently limited. According to the findings presented in the Pulse of the Fashion Industry report, an in-depth analysis of textiles based on fibre type revealed that materials such as leather, silk, and wool exhibit the highest emissions per kilogram of material, whereas polypropylene and acrylic fibres demonstrate the lowest emissions (Huggard, 2020). It is noteworthy that natural fibres like wool, silk, and leather not only contribute significantly to GHG emissions but also have a substantial environmental impact when factors such as water usage and pesticide application during raw material cultivation are considered. Conversely, synthetic fibres, while having lower carbon emissions, are not exempt from environmental concerns. Issues associated with synthetic fibres include their production from non-renewable resources and their contribution to the increasing levels of micro-plastic pollution in the environment. The phase of the value chain that exerts the most influence on GHG emissions varies depending on the fibre type and the specific production processes involved. For instance, a study conducted in the UK highlighted that in the life cycle of a cotton T-shirt, the laundry processes during the use phase consume the highest amount of energy, whereas for a garment made of viscose, the material production phase accounts for the highest energy consumption (Allwood et al., 2015). Consistent with the results of the Pulse of the Fashion Industry report, this study also revealed that garments crafted from non-synthetic fibres consume twice as much energy as those manufactured from synthetic fibres. Furthermore, the emissions linked to different fibres are influenced not only by the type of garments they are transformed into but also by the usage patterns of these garments. Notably, fibres like wool, silk, and leather are commonly utilized in the creation of high-quality, durable garments, while synthetic fibres are predominantly found in fast fashion items.

6 Textile Industry and Methods of Assessment of Carbon Footprint

The assessment of the carbon footprint of the textile industry is imperative for comprehending and addressing its environmental repercussions. Given the intricate nature of textile production processes, selecting an appropriate assessment method necessitates the consideration of factors such as accuracy, comprehensiveness, and practicality (Zhang et al., 2020). A review of the existing literature on carbon footprint evaluation in the textile sector reveals that prevalent evaluation methods encompass ecological cycle assessment, input-output analysis, and mixed life cycle evaluation. Initially, the input-output model serves as an economic quantitative tool that explores the interconnectedness of different components within the economic system throughout the industry cycle. For example, a study was conducted on the influence of international trade on the carbon footprint of American households utilizing the inter-regional input-output analysis model and life cycle assessment, focusing on consumption patterns and expenditure (Weber & Matthews, 2008). Subsequently, another study expanded this research by employing a multi-region input-output model to conduct a comparative analysis across various time frames and household scenarios (Druckman & Jackson, 2009).

The textile industry has a very long life cycle. The textile industry is divided into three phases based on its operations: the first is the agricultural stage, which is the manufacturing and processing of textiles; the second is the industrial stage, which is the cultivation of textile raw materials; and the third is the distribution and transportation of textiles stage or the sales stage. The input-output approach, which is not only targeted but also capable of avoiding truncation errors, is used to analyse the carbon emissions produced by the textile industry's economic activities at various stages.

6.1 Life Cycle Assessment

Life cycle assessment (LCA) is a systematic method used to evaluate the environmental impacts of a product, process, or service throughout its entire life cycle. This includes stages such as raw material extraction, production, transportation, use, and end-of-life disposal or recycling. LCA provides a comprehensive view of the environmental impacts, including carbon emissions, across all stages of a product's life. LCA is used inter alia within industry for optimizing the eco-efficiency of production processes and supply chains, and for making material selection and procurement decisions.

6.1.1 Reasons for Using LCA in the Textile Industry

- Comprehensive Analysis:

Holistic Approach: LCA considers the entire life cycle of textile products, from raw material extraction (e.g., cotton farming, synthetic fibre production) to manufacturing, distribution, use, and end-of-life disposal. This holistic approach ensures that all sources of carbon emissions are accounted for.

Detailed Assessment: LCA allows for detailed quantification of carbon emissions at each stage, providing a clear picture of where the most significant impacts occur.

- Standardization and Consistency:

ISO Standards: LCA is standardized under the ISO 14040 and ISO 14044 guidelines, ensuring consistency and comparability of results across different studies and industries.

Transparency: The standardized methodology provides transparency in the assessment process, making it easier to communicate findings to stakeholders and compare results across studies.

- Identification of Hotspots:

Targeted Mitigation: By identifying the stages in the life cycle where carbon emissions are highest, companies can focus their efforts on these hotspots to achieve the most significant reductions. For example, if the dyeing process is identified as a major contributor, efforts can be made to develop more efficient dyeing techniques or use alternative dyes with lower environmental impacts.

Improvement Opportunities: LCA helps identify opportunities for improvement and innovation in the production process, such as adopting renewable energy sources, improving energy efficiency, or optimizing logistics.

- Data-Driven Decision-Making:

Quantitative Data: LCA provides quantitative data on carbon emissions, which can be used to set measurable reduction targets and track progress over time.

Informed Choices: The detailed data from LCA can inform decisions on material selection, process optimization, and product design to minimize carbon footprints.

- Regulatory and Market Drivers:

Compliance: Many regulations and sustainability standards require companies to report on their environmental impacts. LCA provides a robust method for compliance with these requirements.

Consumer Demand: Increasingly, consumers are demanding transparency and sustainability from brands. LCA allows companies to provide credible and detailed information on their environmental performance, enhancing their reputation and competitive advantage.

- Integration with Other Sustainability Metrics:

Beyond Carbon: While carbon footprint is a critical metric, LCA also evaluates other environmental impacts such as water usage, toxicity, and resource depletion. This comprehensive approach ensures that efforts to reduce carbon emissions do not inadvertently increase other negative impacts.

Balanced Assessment: By integrating multiple sustainability metrics, LCA provides a balanced assessment that supports holistic environmental management.

6.1.2 Challenges and Considerations

While LCA is the most suitable method for calculating the carbon footprint in the textile industry, there are challenges and considerations to keep in mind:

- Data Availability and Quality: High-quality, specific data is required for accurate LCA. Incomplete or inaccurate data can lead to erroneous results.
- Complexity and Cost: LCA can be complex and resource-intensive, requiring expertise and potentially significant investment. However, the long-term benefits in terms of improved sustainability performance often justify these costs.
- Dynamic Nature of the Industry: The textile industry is constantly evolving with new materials, technologies, and practices. LCA studies need to be regularly updated to reflect these changes and provide relevant insights.
- Two terminologies, cradle-to-gate and cradle-to-grave, can explain the life cycle of a product. The cradle-to-gate defines the partial product life cycle, i.e., from resource extraction to factory gate, whereas the cradle-to-grave covers the entire life cycle. Cradle-to-gate mainly focuses on assessing the environmental impacts associated with the production process, such as resource extraction, energy consumption, and emissions. This approach does not consider the product's use phase or end-of-life disposal, making it a valuable tool for identifying and optimising the sustainability of manufacturing processes (Chapman 2010; Zhang et al. 2015). The use phase includes data on the garment lifetime, use frequency, washing habits, and the energy, water and materials used in washing, drying and ironing. In contrast, the end-of-life phase describes emissions from landfills or inputs and emissions from recycling.

An overview of a typical boundary for a "cradle-to-gate" LCA and a "cradle-to-grave" LCA for the textile and garment sector is given in Fig. 2.

The life cycle assessment (LCA) method is the most suitable approach for calculating the carbon footprint of the textile industry due to its comprehensive, standardized, and transparent methodology. It provides detailed insights into the environmental impacts at each stage of the textile product life cycle, enabling targeted and effective mitigation strategies. Despite the challenges associated with

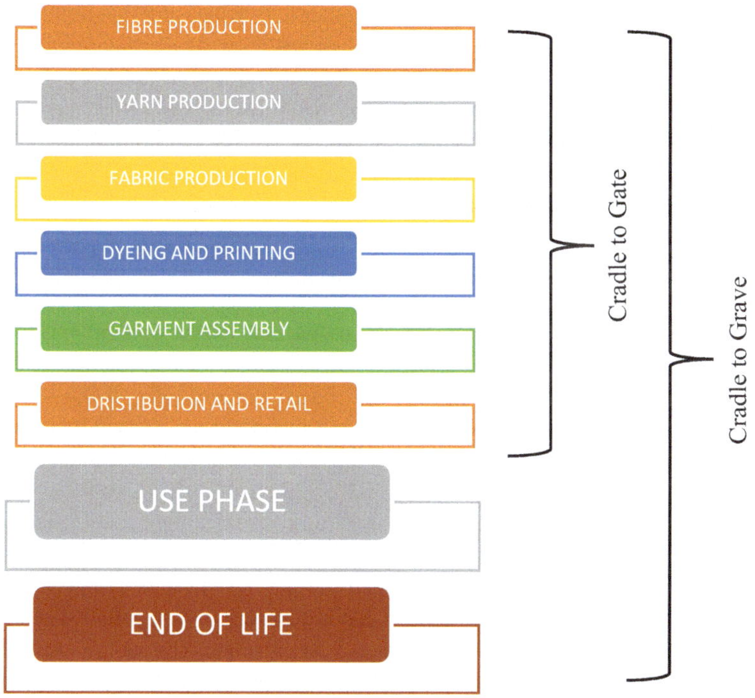

Fig. 2 Overview of boundary for a "cradle-to-gate" LCA and a "cradle-to-grave" LCA

LCA, its benefits in driving data-driven decision-making, regulatory compliance, and sustainability leadership make it the preferred choice for the textile industry.

7 The Process of Calculating Carbon Footprint

Carbon footprinting is essentially a streamlined version of the product environmental footprint (PEF) calculation, both of which are based on ISO 14040, the globally recognized standard for life cycle assessment (LCA). While carbon footprinting focuses exclusively on one impact category—climate change—PEF and LCA take a broader approach. They consider multiple impact categories, including resource use, various environmental impacts, and effects on human health. This comprehensive scope includes factors like energy consumption, habitat impacts, and the release of carcinogens, providing a more holistic view of a product's environmental footprint. ISO 14040 outlines LCA as a four-stage process. The first stage is the goal and scope definition. This initial phase involves clarifying the purpose of the study, identifying the target audience, and setting the boundaries for what will be included in the analysis. For carbon footprinting, an important element of this stage is defining the ‘functional unit’. The functional unit is a quantitative measure that represents the benefits a product is expected to deliver. It serves as a standard reference point, allowing for fair and consistent comparisons between different products. By focusing only on climate change impacts, carbon footprinting simplifies the complex process of LCA. This makes it more accessible and easier to communicate, especially for businesses and consumers looking to understand the climate-related impacts of their products. However, this simplification comes at the cost of not accounting for other significant environmental and health-related impacts. In contrast, PEF and comprehensive LCA provide a more detailed and complete assessment by examining a wide range of impact categories. In carbon footprinting, accurately defining the functional unit is crucial because it underpins the entire assessment. The functional unit allows for a standardized comparison, ensuring that two products are evaluated on an equal basis. For example, if the functional unit is defined as one litre of beverage, all environmental impacts will be calculated per litre, enabling a straightforward comparison between different types of beverages.

Inventory analysis is recognized as the most time-consuming phase in the process of carbon footprinting. This stage involves an in-depth examination of the foreground product system that was established during the goal and scope definition phase. The foreground product system includes all processes directly related to the production, use, and disposal of the product being studied. During inventory analysis, each component of the product’s life cycle is meticulously assessed to identify and quantify potential greenhouse gas emissions. Each step is scrutinized to determine the sources and amounts of greenhouse gases, such as carbon dioxide (CO_2), methane (CH_4), and nitrous oxide (N_2O), that are released as a result of the various processes involved. This detailed examination requires gathering extensive data on inputs and outputs for each life cycle stage. Inputs include raw materials, energy, and other resources required to produce the product, while outputs encompass emissions, waste, and other environmental releases. The data collection process often involves a combination of direct measurements, literature reviews, and modelling, making it a complex and resource-intensive task. The goal of inventory analysis is to create a comprehensive inventory of all relevant greenhouse gas emissions associated with the product system. This inventory serves as the foundation for calculating the product’s overall carbon footprint. By identifying and quantifying emissions at each stage of the life cycle, researchers can pinpoint major sources of greenhouse gases and identify opportunities for emission reductions.

The third stage of carbon footprinting, known as impact assessment, focuses on evaluating the differences between various greenhouse gases and their respective impacts on climate change. This stage is crucial because not all greenhouse gases contribute equally to global warming; they have different properties and effects on the atmosphere’s insulation, as well as varying residence times. In carbon footprinting, the impact assessment commonly considers all greenhouse gases identified by the Intergovernmental Panel on Climate Change (IPCC). These gases include carbon dioxide (CO_2), methane (CH_4), nitrous oxide (N_2O), hydrofluorocarbons (HFCs), perfluorocarbons (PFCs), and sulphur hexafluoride

(SF_6). Each of these gases has a unique global warming potential (GWP), which measures its ability to trap heat in the atmosphere relative to carbon dioxide. Carbon dioxide is the most prevalent greenhouse gas, often used as the baseline for comparison. Methane, although less abundant, has a much higher GWP than carbon dioxide, meaning it is more effective at trapping heat in the short term. Nitrous oxide also has a high GWP and a longer residence time in the atmosphere compared to CO_2. Hydrofluorocarbons, perfluorocarbons, and sulphur hexafluoride, used in various industrial applications, have extremely high GWPs and can remain in the atmosphere for centuries. To standardize the impact assessment, a common approach is to use a 100-year timeframe for analysis. This means calculating the warming potential of each greenhouse gas over a century. This timeframe allows for a balanced view of both short-term and long-term climate effects. By using this standard period, carbon footprinting can provide a more consistent and comparable measure of the different gases' impacts on global warming. During the impact assessment stage, the quantified emissions of each greenhouse gas from the inventory analysis are multiplied by their respective GWP values. This process converts the emissions into a common unit, typically expressed as carbon dioxide equivalents. The carbon dioxide equivalents value represents the total climate impact of all greenhouse gases combined, allowing for a straightforward comparison and summation of their effects.

The interpretation phase of a carbon footprint study is a critical stage where the results from the previous three stages—goal and scope definition, inventory analysis, and impact assessment—are systematically analysed. This phase involves a formal examination of the data and outcomes to ensure they are consistent, reliable, and useful for decision-making. During the interpretation phase, the analyst reviews the entire study to identify any discrepancies or inconsistencies that may have arisen in the earlier stages. This could include errors in data collection, assumptions that need revisiting, or unexpected results from the impact assessment. The cyclical nature of this process means that the analyst may need to go back to previous stages to make corrections or adjustments. For instance, if the inventory analysis reveals unexpected high emissions in a particular life cycle stage, the analyst might return to the goal and scope definition to refine the study boundaries or revisit data sources to ensure accuracy. This iterative approach is valuable because it allows for continuous improvement and refinement of the study. By revisiting earlier stages, the analyst can gain deeper insights into how different decisions and assumptions impact the overall carbon footprint results. This reflective process helps in identifying areas where the product's environmental performance can be enhanced. For example, if the impact assessment shows significant emissions from a specific material, the analyst might explore alternative materials or production methods that could reduce the carbon footprint. The interpretation phase often involves exploring supplementary scenarios to evaluate potential improvements. These scenarios can include changes in materials, production processes, supply chain adjustments, or end-of-life management options. By analysing these alternatives, the analyst can provide recommendations for reducing the carbon footprint and improving the sustainability of the product. By embracing the cyclical nature of this system, analysts can ensure that the carbon footprint study is thorough, accurate, and ultimately beneficial for guiding sustainable decision-making.

7.1 Textile Manufacturing Processes

Textile manufacturing processes involve a series of carefully chosen steps that convert raw materials into finished products, which are typically used for clothing, home furnishing, and various industrial applications. These processes include fibre production, yarn production, and fabric production using techniques like nonwoven, knitting, and weaving, followed by wet treatment, finishing, and sewing. Each of these stages plays a critical role in determining the quality and characteristics of the final textile product. A textile manufacturing facility that manages all these stages from yarn production to the final product is commonly referred to as a 'vertical plant.' This integrated approach allows for greater control over the production process, often resulting in higher efficiency and consistency in the final products. Vertical plants can seamlessly coordinate the various stages, reducing the potential for delays and ensuring that each step is optimized for the next. The raw materials used in textile manufacturing include a diverse range of well-known fibres such as cotton, polyester, viscose, lyocell, wool, and silk. Each material has unique properties that make it suitable for specific applications, and the choice of material impacts the manufacturing process and the environmental footprint of the final product. For instance, cotton is known for its breathability and softness, making it ideal for clothing, while polyester is valued for its durability and resistance to shrinking and stretching, making it suitable for a wide range of applications.

The assessment of the carbon footprint associated with textile manufacturing plants is often conducted within the framework of an Environmental Management System (EMS), an organizational carbon footprint assessment, or a product life cycle assessment (LCA). When the focus is on an EMS or organizational carbon footprint assessment, the functional unit is typically defined as 1 year of operation of the textile plant. This approach provides a comprehensive view of the environmental impact of the plant's operations over time, regardless of the specific processes involved. In contrast, for a product LCA, the assessment is more focused on the specific product being analysed. This requires careful consideration of the distribution methods and the allocation of different processes to the particular product. The LCA aims to provide a detailed understanding of the environmental impacts associated with each stage of the product's life cycle, from raw material extraction to production, use, and disposal. Throughout the textile production processes, there is inevitably some material wastage. This wastage occurs at various stages and must be accounted for in the carbon footprint calculations. Adjustments are made to accommodate these losses from the preceding steps, ensuring that the final carbon footprint reflects the true environmental impact of the manufacturing process. Accounting for material wastage accurately is essential for providing a realistic assessment of the carbon footprint (Collins & Aumonier, 2002).

Therefore, carbon footprint data related to textile processes within the context of a product's carbon footprint cannot be directly translated into carbon footprints of textile manufacturing processes or facilities unless supplementary data outlining the losses at various process stages are provided. The term 'cradle-to-gate' is commonly used in life cycle assessment (LCA) to denote that raw materials are considered in the assessment, while end-use and disposal stages are excluded. This concept is akin to 'cradle-to-plate' utilized in food LCA and differs from 'cradle-to-grave', which entails a comprehensive system analysis.

When exploring guidance on the carbon footprints of textile processes, some numerous studies and reports provide valuable insights. Among these, the IMPRO report authored by Beton et al. in 2014 stands out as a significant resource. Additionally, other noteworthy studies contribute to the understanding of carbon footprints in textile processes. For example, Terinte et al. in 2014 published research on the spin dyeing of modal fibres, which offers detailed information on this specific process and its environmental implications. Another important study is by Muthu et al. in 2012, which focuses on the gains achieved through recycling in the textile industry. Yuan et al. also provided valuable insights in 2012 with their research on pad dyeing, a common technique used in textile manufacturing. An overview provided by Business for Social Responsibility in 2009 offers a broader perspective on the environmental impacts associated with various textile processes. Additionally, the French jeans study provides specific data on the carbon footprint of jeans production, highlighting the environmental challenges and opportunities within this sector. Despite the information contained in these reports, they have not been included in the comparative tables often used in carbon footprint assessments. The primary reason for this exclusion is that the greenhouse gas emissions documented in these studies are typically reported only to the final product. This focus means that the emissions associated with the entire manufacturing process, including any losses that occur during production, are not fully accounted for. To accurately determine the carbon footprint of a manufacturing process or a textile plant, adjustments must be made to include these process losses. Another limitation of these studies is the lack of crucial details necessary for making accurate comparisons among different figures. For instance, specifications on the equipment used in the manufacturing processes are often not provided. This information is essential because different types of equipment can have significantly varying energy efficiencies and environmental impacts. Similarly, the quality of the product outcome, which can affect the overall sustainability of the product, is not always detailed. The type and source of energy used in the processes are also frequently omitted. Since energy sources can greatly influence the carbon footprint, with renewable energy sources generally being more environmentally friendly than fossil fuels, this omission hinders the ability to make precise comparisons.

Notably, recent research from Oakdene Hollins highlighted that primary energy consumption emerges as the most frequently reported impact in life cycle assessments (LCAs) of textiles (Chapman, 2010).

7.2 Textile Product

Expanding on the discussion of textile manufacturing processes outlined in the preceding section, it is imperative to consider the entire life cycle of textile products, encompassing activities related to raw material extraction, distribution, retail, use (including consumer transportation to and from stores and laundry practices), and end-of-life management. Typically, manufacturing and use, characterized by high energy intensity, stand out as the key contributors to the carbon footprint (Roos et al., 2015). Depending on the methodology employed, raw material extraction may also play a significant role in determining the overall carbon footprint of textile products. In spite of research available, there remains a scarcity of published carbon footprinting studies focusing on textile products compared to other categories like foodstuffs or fuels (Peters et al., 2014). Furthermore, the compilation of studies does not encompass research limited to specific parts of the manufacturing process, such as finishing activities. Analysis of the data indicates that the Global Warming Potential over a 100-year period (GWP100) stands out as the primary metric for carbon footprint assessments, yet further methodological intricacies, such as the treatment of biogenic CO_2 emissions and soil carbon variations, or adherence to a standardized carbon footprinting protocol, are seldom clearly articulated in published works. This lack of detailed methodological disclosure likely indicates that existing standards are frequently overlooked and that unconventional methodological aspects are often neglected, including the assumption of climate neutrality regarding biogenic CO_2 emissions.

There exists substantial disparities among apparently similar garments, as evidenced by wool items displaying both the lowest and highest carbon footprints in various studies, ranging from about 8 kg to 46 kg CO_2 equivalents per kilogram of garment. Discrepancies in the calculated carbon footprints mostly stem from factors beyond differences in the employed carbon footprinting methodologies, such as varying assumptions related to user behaviour (e.g., washing temperatures, frequency of washes before disposal) and technological aspects (e.g., manufacturing techniques, electricity sources). An interesting anomaly is observed in the examination of black modal knitted fabrics, which exhibit notably lower carbon footprints per kilogram of fabric compared to other investigations, likely due to the consideration of biogenic carbon stored in the fibres as a carbon sink, resulting in a negative Global Warming Potential (GWP) score. This illustrates the significant impact of the chosen carbon footprint methodology, often overlooked in studies on textile products, particularly due to the presence of bio-based materials. Consequently, life cycle assessment (LCA) practitioners and decision-makers must exercise caution when interpreting carbon footprint outcomes from studies and avoid utilizing them in isolation. To accurately comprehend the results, one must carefully evaluate the specific methodologies and assumptions inherent in each study. This underscores the importance of contextualizing carbon footprint data within the parameters of the research conducted (Peters et al., 2014).

8 Carbon Footprint Labelling of Textile Products

The dissemination of quantitative data through carbon footprint labels is significantly enhancing consumer awareness about the environmental impacts of different products. These labels, which provide specific information about the greenhouse gas emissions associated with a product's entire life cycle, are making consumers more conscious of the ecological footprint of their purchases. The implementation of sophisticated methodologies and approaches for quantifying these emissions has made this possible. As consumers encounter these labels more frequently, they become more informed about the environmental consequences of their consumption choices. This increased awareness influences purchasing decisions, driving a shift towards more environmentally friendly products. When consumers can see the direct impact of their choices in numerical terms, they are more likely to opt for products with lower carbon footprints, thereby supporting sustainable practices. The data provided by these carbon footprint labels do not just inform consumers; they also put pressure on manufacturers and retailers to adopt more sustainable practices. Knowing that consumers are paying attention to these labels, companies are incentivized to reduce their greenhouse gas emissions to appeal to the environmentally conscious market. This can lead to broader industry changes, as businesses innovate and improve their processes to lower their carbon footprints, from sourcing raw materials to final product disposal. The spread of quantitative data through carbon footprint labels is playing a crucial role in elevating consumer awareness about the environmental impacts of product life cycles. By providing clear and precise information on greenhouse gas emissions, these labels empower consumers to make more informed and sustainable choices, while also encouraging businesses to adopt greener practices (Tan et al., 2014).

Numerous labels that incorporate carbon footprint components have been developed for textiles and other commodities, highlighting the increasing focus on sustainability and environmental responsibility. One of the earliest initiatives in this domain was the introduction of the Carbon Reduction Label in the United Kingdom in 2006 by the Carbon Trust. Detailed information about this label and its implementation can be found on the Carbon Trust's website, www.carbontrust.com. This initiative marked a significant step towards providing consumers with transparent information about the carbon emissions associated with their purchases.

When these labels were first introduced, they specifically disclosed the quantity of CO_2 and other greenhouse gases emitted per product or product use. This initial approach aimed to make consumers aware of the environmental impact of their choices in concrete terms. Over time, the Carbon Trust evolved its labelling system to offer two distinct variants: the 'Reducing CO_2' label and the 'CO_2 Measured' label. Each of these labels serves a different purpose in the communication of carbon footprint information. The 'Reducing CO_2' label is designed as a communication tool to demonstrate a company's commitment to reducing its carbon emissions. It not only signifies that a company is taking steps to lower its carbon footprint but may also include a statement regarding the size of the carbon footprint associated with the labelled product or service. Companies that obtain certification for this label are required to commit to reducing their quantified carbon footprint over 2 years. If they fail to achieve this reduction, the label is rescinded. This requirement ensures that companies are actively working towards environmental sustainability rather than merely claiming it. To maintain the integrity and credibility of the 'Reducing CO_2' label, the outcomes of the carbon reduction efforts are evaluated by an independent entity after 2 years. This independent evaluation verifies that the company has appropriately applied the label and has genuinely reduced its carbon footprint as claimed. The rigorous evaluation process underscores the importance of accountability and transparency in environmental labelling, ensuring that the information provided to consumers is accurate and trustworthy.

In contrast to the 'Reducing CO_2' label, the 'CO_2 Measured' label indicates that a product's carbon footprint has been calculated but does not require a commitment to reducing this footprint. The calculation of the carbon footprint follows established standards such as PAS2050 or the GHG Protocol. These standards ensure a comprehensive evaluation that spans the entire life cycle of the product, from production to disposal. This label serves to provide transparency about the environmental impact of a product, allowing consumers to make informed decisions even if the manufacturer has not yet committed to reducing emissions. Another significant company in the realm of carbon footprint labelling is Carbon Care Asia, a company based in Hong Kong. Carbon Care Asia has developed a labelling system that assesses the efforts of companies and organizations in reducing their carbon footprint. Their approach involves three critical stages: measuring, reducing, and offsetting carbon footprints. This structured methodology ensures that companies systematically address their emissions and work towards meaningful reductions. Prospective users of the Carbon Care label must first provide a verified audit report of their emission inventory, conducted according to the Carbon Care label protocol. This verification process ensures the accuracy and credibility of the reported emissions. Based on the extent of emission reductions achieved compared to baseline emissions, six different labelling options are available. These labels reflect the level of accomplishment in carbon reduction, with categories indicating reductions ranging from 5% to 80% relative to the measured emission baselines. For entities that achieve carbon neutrality, a prestigious golden label is awarded. This accolade signifies that the organization has not only measured and reduced its emissions but has also offset any remaining emissions, effectively achieving a net-zero carbon footprint. The golden label is a mark of exceptional commitment to environmental sustainability, distinguishing those entities that go above and beyond in their carbon management efforts (CarbonCareAsia, 2014).

An American charitable organization called Carbonfund.org created the CarbonFree Product Certification to inform consumers about companies taking responsibility up for their greenhouse gas emissions. This certification covers a wide range of greenhouse gases, including methane (CH_4), nitrous oxide (N_2O), sulphur hexafluoride (SF_6), hydrofluorocarbons (HFCs), perfluorocarbons (PFCs), and carbon dioxide (CO_2) emissions from biomass. By doing so, the certification aims to highlight companies that are committed to mitigating their environmental impact. To achieve the CarbonFree certification, a comprehensive life cycle assessment (LCA) must be conducted to calculate the greenhouse gas emissions produced throughout the entire life cycle of a product. This life cycle spans from raw material extraction to production, distribution, use, and disposal. The assessment ensures that all stages of the product's life are considered, providing a complete picture of its environmental footprint. Once the product's life cycle carbon footprint has been quantified, any emissions that cannot be reduced are offset through independently verified carbon offset programmes. These programmes focus on areas such as energy efficiency, forestry, and renewable energy. By participating in these offset programmes, companies can effectively neutralize their remaining emissions, contributing to broader environmental sustainability efforts. The methodologies used for measuring greenhouse gas emissions under the CarbonFree certification are grounded in internationally recognized standards. These include the GHG Protocol, PAS 2050, and ISO 14044. The GHG Protocol provides a comprehensive global framework for measuring and managing greenhouse gas emissions. PAS 2050 offers a specific standard for assessing the life cycle greenhouse gas emissions of goods and services, while ISO 14044 specifies requirements and guidelines for life cycle assessments. By adhering to these standards, the CarbonFree certification ensures that the measurement and offsetting processes are robust, credible, and transparent (Carbonfund, 2022).

9 Examining Issues with Textile Industry Carbon Footprint Study

The carbon footprint accounting for textile products is fraught with uncertainty and incompatibility due to the intricate and extensive nature of their life cycles. This complexity begins with agricultural processes, where raw materials like cotton are grown and extends through industrial processes where textiles are manufactured. It also encompasses the entire period of the product's use by consumers and culminates in the final stages of the product's life, such as disposal or recycling. Each phase of this extensive life cycle contributes to the overall carbon footprint, making it difficult to maintain high-quality data throughout the entire process.

One significant challenge in carbon footprint accounting for textiles is the lack of detailed explanations provided by researchers on their methodologies. Many studies fail to thoroughly explain their accounting processes, including how they distribute carbon emissions across different stages of the life cycle and the principles they use to make trade-offs between various environmental impacts. This lack of transparency hampers the ability to trace and verify the rationality of their carbon footprint results. Without a clear understanding of these underlying methodologies, it becomes difficult to assess the accuracy and reliability of the reported carbon footprints. Additionally, discrepancies arise from the use of different accounting models and boundaries by various researchers. These models and boundaries define the scope of what is included in the carbon footprint calculations, and variations in these parameters can lead to vastly different results. For instance, one study might consider only direct emissions from manufacturing, while another might include indirect emissions from transportation and raw material production. Such inconsistencies make it impossible to conduct meaningful horizontal comparisons of carbon footprint results across different studies. As a result, stakeholders, including policymakers, businesses, and consumers, face significant challenges in making informed decisions based on these inconsistent and incomparable data. The complexity of the textile product life cycle, combined with the lack of detailed methodological explanations and the use of varying accounting models, significantly impedes the accuracy, traceability, and comparability of carbon footprint results. Addressing these issues requires greater transparency and standardization in carbon footprint accounting practices to ensure reliable and comparable data for effective decision-making in the textile industry.

The analysis and division of carbon footprint stages for textile products have garnered increasing attention across different fields and industries. However, within the domestic research context, there has been a relatively limited focus specifically on the textile industry. Despite this growing interest, existing studies often do not delve deeply into the unique complexities and specificities of textiles when it comes to carbon footprint analysis. Most of the literature available applies life cycle theory as a framework to estimate carbon emissions associated with various stages of textile products' life cycles. These stages typically include design, production, sales, and recovery. Life cycle theory is a comprehensive approach that considers the environmental impacts associated with all phases of a product's life, from raw material extraction through manufacturing, usage, and end-of-life disposal or recycling. By using this theory, researchers aim to capture the full spectrum of carbon emissions attributable to textile products. However, there is often a lack of comprehensive analysis concerning the sources of carbon footprints throughout the entire life cycle of textile products. Many studies focus on specific stages without integrating a holistic view of how emissions are generated and accumulated from one stage to the next. This piecemeal approach can overlook significant interactions and dependencies between stages, leading to an incomplete understanding of the total carbon footprint. Moreover, the establishment of clear and consistent accounting boundaries is frequently missing in most studies. Accounting boundaries define what is included and excluded in carbon footprint calculations, and inconsistency in these boundaries can lead to significant variations in reported results. For instance, one study might only consider emissions from direct production processes, while another might include emissions from upstream activities such as raw material extraction and transportation. Without standardized boundaries, comparisons between studies become challenging, and the overall reliability of the carbon footprint data is compromised. To enhance the accuracy and comparability of carbon footprint assessments in the textile industry, there is a need for more detailed, holistic analyses and standardized methodologies.

One of the key shortcomings of current carbon footprint accounting methods is their narrow scope. While they effectively quantify emissions from production processes, they do not provide a comprehensive evaluation of how these emissions influence environmental and economic development. For instance, these methods do not adequately address how carbon emissions affect resource depletion, biodiversity loss, or social issues such as health and employment. Consequently, the results generated by these accounting methods offer limited guidance for developing strategies aimed at achieving sustainable development in a holistic manner. Moreover, the ability to compare the carbon footprints of identical textile products across different regions holds significant potential. Such comparisons allow for a more nuanced understanding of how the environmental and social impacts of production vary by location. For example, the carbon footprint of a textile product manufactured in one region might be significantly different from the same product produced in another region due to variations in energy sources, production technologies, and regulatory frameworks. By quantifying these differences, researchers and policymakers can gain valuable insights into the specific environmental and social challenges faced by each region. This comparative information is crucial for formulating targeted regional measures to reduce energy consumption in production processes. If a particular region is identified as having a higher carbon footprint for certain textile products, specific interventions can be designed to address the underlying causes. These might include investing in cleaner energy sources, enhancing production efficiency, or implementing stricter environmental regulations. By tailoring measures to the specific needs and conditions of each region, it is possible to achieve more effective and sustainable reductions in carbon emissions (Zhang et al., 2020).

10 Step Towards Mitigating Carbon Emissions and Achieving UNSDGs

The fashion industry is a significant contributor to greenhouse gas (GHG) emissions, and to mitigate its impact on climate change, a shift from a moderate decarbonization trajectory to a more aggressive approach is necessary (Berg & Karl-Hendrik, 2021). This shift is crucial for the industry to align with the goals of the Paris Climate Change Agreement, which aims to limit global warming to 1.5 °C above pre-Industrial Revolution levels (UNFCCC, 2018). However, achieving these goals is challenging due to globalization, the drive for economic growth, and the lack of effective global policies. Large-scale systemic change is essential for developing a sustainable fashion system, which requires collaboration among stakeholders, technological innovation, supportive government policies, and adequate infrastructure.

Systemic Change Requirements

1. Stakeholder Collaboration: All stakeholders, including brands, suppliers, consumers, and policymakers, need to work together to drive systemic change. Collaborative efforts are essential to create a unified approach to sustainability.

2. Technology Innovation: Advancements in technology are crucial for developing sustainable solutions. Innovations in farming, recycling, and biomaterials can significantly reduce the industry's environmental impact.
3. Government Policy and Infrastructure Support: Effective government policies and infrastructure support are necessary to facilitate the transition to a sustainable fashion system. Policies should focus on promoting renewable energy, providing financial incentives, and removing barriers to decarbonization.

Sustainable Solutions

1. Technical Advancements: Implementing technical advancements in farming, such as regenerative agriculture, can help reduce the environmental impact of raw material production.
2. Recycling Innovations and Biomaterials: Innovations in recycling and the development of biomaterials can contribute to a circular economy by reducing waste and promoting the reuse of materials.
3. Reuse and Resale Initiatives: Encouraging reuse and resale initiatives can support a circular economy and diminish the industry's ecological footprint.

Decarbonization Strategies

1. Phasing Out Coal: Transitioning from coal to renewable energy sources is essential for reducing GHG emissions in the fashion industry (Chen et al., 2021).
2. Bio-Based Feedstock: Using bio-based feedstock can help reduce the carbon footprint of textile production.
3. Low-Emission Heat Sources and Sustainable Production: Implementing low-emission heat sources and sustainable production processes can further reduce the industry's environmental impact.
4. Increased Recycling Rates: Increasing the rate of recycling materials is crucial for transitioning to a circular business model.
5. Regeneration of Natural Systems: Promoting the regeneration of natural systems can help restore ecosystems and enhance sustainability (Jia et al., 2020).

Government Policy and Financial Support

1. Manufacturing Hub Policies: Policies targeting manufacturing hubs are critical for removing impediments to decarbonization.
2. Financial Incentives: Financial support through subsidies, incentives, and lower loan rates can encourage investments in renewable energy and sustainable practices.
3. Infrastructure Development: Developing infrastructure for renewable energy sources is essential for supporting the industry's transition to sustainability.

Transformative Policies and Regulations

1. Tighter Supply Chain Obligations: Implementing stricter supply chain obligations and incentives can promote better business models (Villena & Gioia, 2020).
2. Extended Producer Responsibility (EPR) Legislation: EPR legislation can fund the collection, sorting, and recycling of clothing at the end of its life cycle, reducing waste and promoting sustainability (Bukhari et al., 2018).

Retailer-Supplier Partnerships
Stable partnerships between retailers and suppliers are needed to enable long-term investments and share the costs of cleaner and greener production methods. These partnerships can foster mutual commitment to sustainability goals and ensure the industry's transition to more sustainable practices.

Consumer Behavior and Aesthetic Requirements
Consumer preferences are evolving, with increased demand for products that are both functional and attractive, as well as environmentally friendly. Textiles known for their environmental benefits, or those marketed as such, are expected to play a more prominent role in consumer choices. The fashion industry must shift from its focus on low costs, high volumes, and constant novelty to a growth paradigm that prioritizes sustainability. Marketing messages need to promote sustainable consumption behaviours, including:

1. Preference for Sustainable Materials: Encouraging the use of sustainable materials in fashion products
2. Environmentally Friendly Washing and Drying Methods: Promoting garment care practices that reduce environmental impact
3. Garment Recycling, Rental, and Repair: Supporting practices that extend the life of garments through recycling, rental, and repair
4. Reuse and Sharing Options: Encouraging consumers to embrace reuse and sharing of clothing to reduce waste (Niinimäki et al., 2020; Berg & Karl-Hendrik, 2021)

The fashion industry must undergo a significant transformation to address its environmental impact. By adopting sustainable practices, collaborating with stakeholders, and implementing supportive policies, the industry can move towards a more sustainable future, ultimately contributing to global efforts to combat climate change.

11 Concepts for Future Research on the Textile Industry's Carbon Footprint

The textile industry, one of the largest global contributors to environmental degradation, is under increasing scrutiny for its substantial carbon footprint. As the industry continues to grow, driven by fast fashion and increasing consumer demand, the need for sustainable practices becomes more pressing. Future research must delve into innovative solutions and strategies to mitigate these emissions, focusing on the adoption of eco-friendly materials, advancements in manufacturing technologies, and the implementation of circular economy principles. Understanding the complex interplay between industry practices and their environmental impact is crucial for developing effective policies and fostering a more sustainable future for the textile sector.

Establishing a standardized carbon footprint assessment method based on life cycle principles for the textile industry is imperative to ensure consistency, accuracy, and comparability in measuring environmental impacts. Such a method would provide a comprehensive evaluation of the carbon emissions associated with each stage of a textile product's life cycle, from raw material extraction through production, distribution, use, and disposal. By adopting standardized metrics and methodologies, stakeholders across the industry—including manufacturers, policymakers, and consumers—can make informed decisions that drive sustainability. This approach would also facilitate transparency, enabling more effective regulation and fostering innovation in low-carbon technologies and practices within the textile sector.

Analysing the temporal and spatial distribution of the textile industry throughout its entire life cycle is crucial for understanding its full environmental impact. This analysis involves examining how carbon emissions and other environmental effects vary over time and across different geographic locations, from the sourcing of raw materials to manufacturing, distribution, use, and disposal. By identifying patterns and hotspots of high emissions or resource use, stakeholders can target interventions more effectively, optimizing processes, reducing waste, and minimizing the industry's overall carbon footprint. Such insights are essential for developing tailored, region-specific strategies that address both global and local sustainability challenges, ensuring a more sustainable future for the textile industry.

Incorporating a carbon footprint impact assessment into carbon footprint accounting is essential for providing a comprehensive understanding of the environmental and societal effects of carbon emissions. While carbon footprint accounting quantifies the amount of greenhouse gases produced by various activities, an impact assessment goes further by evaluating the broader implications of these emissions on ecosystems, public health, and social well-being. This holistic approach allows stakeholders to not only track and reduce emissions but also to understand and mitigate their consequences. By quantifying the environmental and societal impacts, decision-makers can prioritize actions that promote sustainable development, enhance quality of life, and support the transition to a low-carbon economy. This addition ensures that carbon management strategies are aligned with broader environmental and social objectives, leading to more effective and equitable outcomes.

References

Advanced Manufacturing Office. (n.d.). Energy.gov. Retrieved June 29, 2024, from http://www1.eere.energy.gov/industry/industries_technologies/index.html

Alkaya, E., & Demirer, G. N. (2014). Sustainable textile production: A case study from a woven fabric manufacturing mill in Turkey. *Journal of Cleaner Production, 65*, 595–603.

Allwood, J. M., Laursen, S. E., de Rodriguez, C. M., & Bocken, N. M. (2015). Well dressed?: The present and future sustainability of clothing and textiles in the United Kingdom. *Journal of the Home Economics Institute of Australia, 22*(1), 42.

Berg, A., & Karl-Hendrik, M. (2021). *Fashion on climate—How the fashion industry can urgently act to reduce its greenhouse gas emissions*. McKinsey & Company and Global Fashion Agenda.
Beton, A., Dias, D., Farrant, L., Gibon, T., Le Guern, Y., Desaxce, M., Perwueltz, A., Boufateh, I., Wolf, O., Kougoulis, J., & Cordella, M. (2014). Environmental improvement potential of textiles (IMPRO textiles). *European Commission, 20*.
Bevilacqua, M., Ciarapica, F. E., Giacchetta, G., Bevilacqua, M., Ciarapica, F. E., & Giacchetta, G. (2012). Case study: A carbon footprint analysis in textile supply chain. In *Design for environment as a tool for the development of a sustainable supply chain* (pp. 219–239).
Biswas, I., Raj, A., & Srivastava, S. K. (2018). Supply chain channel coordination with triple bottom line approach. *Transportation Research Part E: Logistics and Transportation Review, 115*, 213–226.
Bukhari, M. A., Carrasco-Gallego, R., & Ponce-Cueto, E. (2018). Developing a national programme for textiles and clothing recovery. *Waste Management & Research, 36*(4), 321–331.
Cai, Y. J., & Choi, T. M. (2020). A United Nations' sustainable development goals perspective for sustainable textile and apparel supply chain management. *Transportation Research Part E: Logistics and Transportation Review, 141*, 102010.
Carbonfree *Product Certification Carbon Footprint Protocol*. (2022). https://carbonfund.org/wp-content/uploads/2022/12/Carbonfree-Product-Certification-Protocol-v8.0-112722.pdf
Chapman, A. (2010). *Mistra future fashion–review of life cycle assessments of clothing*. Oakdene Hollins Research & Consulting.
Chen, X., Memon, H. A., Wang, Y., Marriam, I., & Tebyetekerwa, M. (2021). Circular economy and sustainability of the clothing and textile industry. *Materials Circular Economy, 3*, 1–9.
Collins, M., & Aumônier, S. (2002). *Streamlined life cycle assessment of two Marks & Spencer plc apparel products*. Marks & Spencer plc.
Dahlbo, H., Aalto, K., Eskelinen, H., & Salmenperä, H. (2017). Increasing textile circulation—Consequences and requirements. *Sustainable Production and Consumption, 9*, 44–57.
Dissanayake, G., & Sinha, P. (2015). An examination of the product development process for fashion remanufacturing. *Resources, Conservation and Recycling, 104*, 94–102.
Druckman, A., & Jackson, T. (2009). The carbon footprint of UK households 1990–2004: A socio-economically disaggregated, Quasi-multi-regional input–output model. *Ecological Economics, 68*(7), 2066–2077.
Finkbeiner, M. (2009). Carbon footprinting—Opportunities and threats. *The International Journal of Life Cycle Assessment, 14*, 91–94.
Gereffi, G. (2001). Beyond the producer-driven/buyer-driven dichotomy the evolution of global value chains in the internet era. *IDS Bulletin, 32*(3), 30–40.
Halldórsson, Á., Kotzab, H., & Skjøtt-Larsen, T. (2009). Supply chain management on the crossroad to sustainability: A blessing or a curse? *Logistics Research, 1*, 83–94.
Hasan, A. M., Rokonuzzaman, M., Tuhin, R. A., Salimullah, S. M., Ullah, M., Sakib, T. H., & Thollander, P. (2019). Drivers and barriers to industrial energy efficiency in textile industries of Bangladesh. *Energies, 12*(9), 1775.
Hasanbeigi, A. (2010). *Energy-efficiency improvement opportunities for the textile industry*. Lawrence Berkeley National Laboratory.
Hasanbeigi, A., & Price, L. (2015). A technical review of emerging technologies for energy and water efficiency and pollution reduction in the textile industry. *Journal of Cleaner Production, 95*, 30–44.
Huggard, E. (2020). The fashion industry context. In *Communicating fashion brands* (pp. 20–38). Routledge.
IPCC. (2007). AR4 climate change 2007: The physical science basis — IPCC. Ipcc.ch; IPCC. https://www.ipcc.ch/report/ar4/wg1/
Jia, F., Yin, S., Chen, L., & Chen, X. (2020). The circular economy in the textile and apparel industry: A systematic literature review. *Journal of Cleaner Production, 259*, 120728.
Kirchain, R., Olivetti, E., Miller, T. R., & Greene, S. (2015). *Sustainable apparel materials* (pp. 1–34). Materials Systems Laboratory, Massachusetts Institute of Technology.

Leal Filho, W., Perry, P., Heim, H., Dinis, M. A. P., Moda, H., Ebhuoma, E., & Paço, A. (2022). An overview of the contribution of the textiles sector to climate change. *Frontiers in Environmental Science, 10*, 973102.

Masson-Delmotte, V., Zhai, P., Pirani, A., Connors, S. L., Péan, C., Berger, S., Caud, N., Chen, Y., Goldfarb, L., Gomis, M. I., & Huang, M. (2021). Climate change 2021: The physical science basis. *Contribution of Working Group I to the Sixth Assessment Report of the Intergovernmental Panel on Climate Change, 2*(1), 2391.

McCullough, D. G. (2014). Deforestation for fashion: Getting unsustainable fabrics out of the closet. *The Guardian, 25*.

Muthu, S. S., Li, Y., Hu, J. Y., & Ze, L. (2012). Carbon footprint reduction in the textile process chain: Recycling of textile materials. *Fibers and Polymers, 13*(8), 1065e1070.

Niinimäki, K., Peters, G., Dahlbo, H., Perry, P., Rissanen, T., & Gwilt, A. (2020). The environmental price of fast fashion. *Nature Reviews Earth & Environment, 1*(4), 189–200.

Peters, G. P., & Hertwich, E. G. (2008). Post-Kyoto greenhouse gas inventories: Production versus consumption. *Climatic Change, 86*(1), 51–66.

Peters, G., Granberg, H., & Sweet, S. (2014). The role of science and technology in sustainable fashion. In *Routledge handbook of sustainability and fashion* (pp. 181–190). Routledge.

Peters, G., Svanström, M., Roos, S., Sandin, G., & Zamani, B. (2015). Carbon footprints in the textile industry. In *Handbook of life cycle assessment (LCA) of textiles and clothing* (pp. 3–30). Woodhead Publishing.

Quantis. (2018). Measuring Fashion 2018. *Environmental impact of the global apparel and footwear industries study full report and methodological considerations*.

Roos, S., Sandin, G., Zamani, B., & Peters, G. (2015). Environmental assessment of Swedish fashion consumption. In *Five garments–sustainable futures*. Mistra Future Fashion.

Sandin, G., Roos, S., Spak, B., Zamani, B., & Peters, G. (2019). *Environmental assessment of Swedish clothing consumption—Six garments* (p. 6). Sustainable Futures.

Second Briefing Session held for CarbonCare® Label 2014 and ESG Reporting – Climate Finance Asia. (2014). Retrieved July 1, 2024, from https://www.climatefinanceasia.com/events/second-briefing-session-held-for-carboncare-label-2014-and-esg-reporting/

Stoy, P. C. (2018). Deforestation intensifies hot days. *Nature Climate Change, 8*(5), 366–368.

Tan, M. Q. B., Tan, R. B. H., & Khoo, H. H. (2014). Prospects of carbon labelling–a life cycle point of view. *Journal of Cleaner Production, 72*, 76–88.

Terinte, N., Manda, B. K., Taylor, J., Schuster, K. C., & Patel, M. K. (2014). Environmental assessment of coloured fabrics and opportunities for value creation: Spin-dyeing versus conventional dyeing of modal fabrics. *Journal of Cleaner Production, 72*, 127–138.

United Nations. (2018, September 6). *UN Helps Fashion Industry Shift to Low Carbon*. United Nations Climate Change. https://unfccc.int/news/un-helps-fashion-industry-shift-to-low-carbon

Villena, V. H., & Gioia, D. A. (2020). A more sustainable supply chain. *Harvard Business Review, 98*(2), 84–93.

Viscose *Staple Fibre – Market & Industry Manufacturers*. (2024). Www.mordorintelligence.com. https://www.mordorintelligence.com/industry-reports/viscose-staple-fiber-market

Weber, C. L., & Matthews, H. S. (2008). Quantifying the global and distributional aspects of American household carbon footprint. *Ecological Economics, 66*(2–3), 379–391.

Wiedmann, T., & Minx, J. (2008). A definition of 'carbon footprint'. *Ecological Economics Research Trends, 1*(2008), 1–11.

World Population Prospects – Population Division – United Nations. (n.d.). Population.un.org. Retrieved June 29, 2024, from http://esa.un.org/unpd/wpp/unpp/panel_population.htm

Yasin, S., Behary, N., Rovero, G., & Kumar, V. (2016). Statistical analysis of use-phase energy consumption of textile products. *The International Journal of Life Cycle Assessment, 21*, 1776–1788.

Zamani, B., Sandin, G., & Peters, G. M. (2017). Life cycle assessment of clothing libraries: can collaborative consumption reduce the environmental impact of fast fashion?. *Journal of cleaner production, 162*, 1368–1375.

Zhang, X., & Chen, Y. (2019). Carbon emission evaluation based on multi-objective balance of sewing assembly line in apparel industry. *Energies, 12*(14), 2783.

Zhang, Y., Liu, X., Xiao, R., & Yuan, Z. (2015). Life cycle assessment of cotton T-shirts in China. *The International Journal of Life Cycle Assessment, 20*, 994–1004.

Zhang, J., Qian, X., & Feng, J. (2020). Review of carbon footprint assessment in textile industry. *Ecofeminism and Climate Change, 1*(1), 51–56.

Aligning Apparel and Textile Sector Practices with the United Nations' Sustainable Development Goals: Balancing the Pillars of Sustainability

R. Surjit, V. Sruthi, K. S. Tarunvalavan, and H. Shanruthi

1 Introduction

The United Nations, an international organisation, has created its 2030 Agenda for Sustainable Development, in the year 2015. The agenda majorly included 17 Sustainable Development Goals (SDGs), also called the global goals. The agenda had a motive to create, and to bring about prosperity and peace, not only among the people but the planet as well, while at the same time preserving the natural environment and dealing with climatic changes. This chapter deals with three main pillars that are associated with these sustainable development goals. It includes the major environmental, social, and economic factors. However, the idea of sustainability remains the central or core idea of the UN's vision of SDGs. All the other aspects and factors revolve around this central ideology.

With the ever-growing demand for sustainable products in recent times, the steps taken by every sector to produce eco-friendly goods have grown eventually. Moreover, the measures taken by the apparel and textile sectors towards achieving it are immense. These two sectors have made significant contributions towards the above-mentioned UN SDGs, which have also been discussed in detail in this chapter. It includes significant advancements in the apparel and textile supply chains. Along with the existence of the three pillars of sustainability, the practicality of implementing and achieving the same should also be taken into consideration. UN's goal can be achieved to its fullest only when the three sustainability pillars are concentrated equally and balanced properly. Balancing, on a brief note involves

R. Surjit (✉) · V. Sruthi · K. S. Tarunvalavan · H. Shanruthi
Department of Fashion Technology, PSG College of Technology, Coimbatore, Tamil Nadu, India
e-mail: rst.fashion@psgtech.ac.in

S. S. Muthu (ed.), *Sustainable Textile and Apparel Chain Management*, Sustainable Textiles: Production, Processing, Manufacturing & Chemistry,
https://doi.org/10.1007/978-3-031-80240-9_9

the balance in terms of economic growth, environmental protection, social equity, and the interconnectedness associated with the same.

Not all company or a firm is run by a single person, or at least the decisions in it cannot be taken all by themselves. With the common presence of shareholders or stakeholders in most industries and factories, their cooperation and willingness to work towards sustainable development matter a lot. Moreover, these stakeholders play a crucial role in bringing out and ensuring a proper balance of the three pillars of sustainability. Though the practicality of taking the initiative to bring about these practices comes with a lot of obstacles and hurdles, there is considerable strategic planning that can be done. This will also be discussed in this chapter. However, in the long term, the previously defined goals and strategies will all be upgraded to a large extent, giving hope for further research in the future.

2 Pillars of Sustainability and UN SDGs

In the textile and apparel industries, sustainability has been the buzzword and has created paramount importance over the last 15 years. The recent years have witnessed a great increase in the publications on the topics relevant to 'sustainability'. The paradigm 'pillars of sustainability' arose due to a complicated supply chain in many industries including the textile and apparel sectors. The study of the three pillars role in textile and apparel should be done due to the lack of theoretical analysis (Boussemart et al., 2020). This chapter will give details and enterprising solutions for sustainable and sophisticated apparel processes. Generally referred to as the major three pillars that hold and support sustainability, the following are the so-called 'intertwined pillars' of sustainability:

(i) Environmental sustainability
(ii) Social sustainability
(iii) Economical sustainability

They are called so because they exhibit potential interdependence between them. From the viewpoint of the globalisation process, we always aim at sustainable development. The pillars are 'intertwined' in terms of global energy system (Duic et al., 2015). There are possible spillover effects of every action taken on one's side on other pillars of sustainability. For instance, ideal resource management can improve a country's economy. The quality of the environment decides the health of life on earth; therefore, if people are healthy, the economy would be increasing. There is always a coherence between each pillar so that they are interconnected to make a sustainable society of healthy people, environment, and economy. Purvis et al. (2019) provided illustrations like the Venn diagram, concentric circle, and pillar structure as the ideal approaches to understanding sustainability. The work stated that sustainable goals can be achieved only when they are balanced. So there is a need to balance the pillars to shape a sustainable future in the textile world. Despite contemporary literature which has a perception that sustainability revolves around the UN SDGs, the pillars play a critical role in being implicitly embedded in their certain goals (UN, 2012).

2.1 Environmental Sustainability

Environmental sustainability is the ability to maintain the desired properties or features of our physical environment without having to compromise on the needs of future generations. Environmental sustainability due to its vital imperativeness is considered the most critical pillar of all pillars as this type affects the other two. It is accountable to conserve the natural environment and shelter the ecosystem globally to ensure the healthy well-being of the present and future of our peer groups. Apparel and textile companies unquestionably have a responsibility to ensure environmental sustainability. With much clarity, the stakeholders must implement eco-friendly practices within their apparel and textile firms. The major problem in developing such practices is that the ongoing sustainable behaviours and efforts are completely differentiated from the companies' business targets in terms of the perception of the stakeholders. The sustainable goals should be aligned with the business goals and profit for an efficient implementation of environmental sustainability in the firm. A lot of influencing factors such as increased population growth result in higher greenhouse gas emissions and deforestation. Our planet Earth is reaching the threshold point. As a result, one can visualise the environmental consequences of global warming all around the globe. For a liveable future, every apparel and textile business should invest in environmentally sustainable energies and technologies.

2.2 Economical Sustainability

Economical sustainability ensures that businesses all around the world improve the standard of living of all the labour, staff, and stakeholders analogous to the business.

It focuses on two sets of principles. They are:

1. *Business practices aimed to achieve ideal economic growth*
2. *Productive utilisation of the company's wealth to promise fair profitability over the period*

It mainly concentrates on avoiding short-term processes and aims at the globe's long-term sustainable development. A sustainable economy has paramount importance for several reasons and we could prove it to be reasonable as this directly influences environmental and social sustainability. The economy of a country is based on its people. Economic parameters that tell about the human well-being in the world such as poverty, hunger, life expectancy, and literacy rates are considered while taking economic sustainability as a concern. In the past years, many textile and apparel companies have been reporting 'low wages' paid to the workers. 'The global natural resources are non-renewable resources'. As the quote says, natural resources are very limited; the endurance of the global economy is essential for the long run. It is majorly done for the preservation of human life and in need to save the planet for the upcoming generation. Economic sustainability can be achieved in many forms. For instance, businesses invest in sustainable technologies and products to make the environment better for the people. Converting business operations into renewable sources of energy is also one of the possible ways to travel in the long journey (Masterclass, 2022).

2.3 Social Sustainability

Even though there is widespread usage of the word '*sustainability*', people often find it difficult to define the term due to its varied concepts. Davidson (2010) has reviewed how the social element of sustainability has played a vital role in the context of both academic and policy debate. The definition of social sustainability is *the development (and/or growth) that is compatible with the harmonious evolution of civil society, fostering an environment conducive to the compatible cohabitation of culturally and socially diverse groups while at the same time encouraging social integration, with improvements in the quality of life for all segments of the population* (Stren & Polèse, 2000).

There are totally five major principles included in terms of this type of sustainability.

Five major principles constituting social sustainability are given below:

(i) Inclusion and diversity
(ii) Social equity
(iii) Democracy and empowerment of people
(iv) Good quality of life
(v) Livelihood security

Despite the principles, the key factors affecting the implementation of this social sustainability are inclusion and social equity. Community engagement is imperative and access to resources and services is basic. Social sustainability is essential to rescue the future and initiate a fair and justifiable treatment of each other socially. Therefore, this social element is also considered an important part when concerning pillars of sustainability. The UN SDGs are to remove poverty, ensure good health and well-being of people, eradicate hunger, provide quality education and gender equality, provide pure water and better sanitation facilities, improve prospects of acquiring affordable and clean energy, provide decent work and aid in economic growth, provide industrial growth, innovation, and infrastructure, reduce inequalities, have cities and communities that are sustainable, ensure responsibility in production and consumption, take action on climate issues, take care of life below

Table 1 Environmental element of sustainability (Dalampira & Nastis, 2020)

Group	Title of SDGs	Description of SDGs
Environmental sustainability	*Clean water and sanitation*	Ensure water sanitation
	Affordable and clean energy	Promote access to cheap and reliable sources of sustainable energy
	Life below water	Preserving all water bodies and sustainable use of those resources
	Climate action	Initiatives to eliminate climatic change and its after effects
	Life on land	Arrest deforestation and end biodiversity loss and sustainable management of forests

Table 2 Social element of sustainability (Dalampira & Nastis, 2020)

Group	Title of SDGs	Description of SDGs
Social sustainability	*Reduce inequalities*	Abolish inequality among countries
	Gender equality	End gender inequality and strengthen women's empowerment
	Peace, justice, and strong institutions	Promote inclusive and peaceful societies
	Sustainable cities and communities	Ensure safe and sustainable cities
	Quality education	Promote and ensure quality education for all
	Good health and well-being	Promote good well-being of people

Table 3 Economical element of sustainability (Dalampira & Nastis, 2020)

Group	Title of SDGs	Description of SDGs
Economic sustainability	*No poverty*	Abolish inequality among countries
	Responsible consumption and production	Ensure the use of resources and sustainable production
	Industry, innovation, and infrastructure	Build robust infrastructure and promote innovation for sustainable industrialisation
	Zero hunger	Eradicate hunger and promote agriculture and nutrition
	Decent work and economic growth	Promote decent work for all and hence economic growth
	Partnerships for the goals	Strengthen partnerships for sustainable development on a global level

water and on land, ensure peace and justice, and create strong institutions and forge partnerships for the goals (Bain et al., 2019). Tables 1, 2, and 3 group the UN Sustainable Development Goals based on the three types of pillars of sustainability:

3 Contribution of Apparel and Textile Sectors Towards UN SDGs

In the apparel and textile sectors, sustainability is the major growing concern in all the ways possible, that is, to preserve the environment for future generations. The conventional fashion industry is one of the most polluting industries as it undergoes a lot of wet processing in the production stage, carbon emission in the supply chain, and finally ending up in landfills. Now, the polluting index is reduced to some extent by following the United Nations Sustainable Development Goals. The major contributions of the apparel and textile sectors towards the goals are as follows:

3.1 SDG 1: No Poverty

The apparel and textile industries are the major job-providing industries, especially in developing countries, which helps them reduce the unemployment rate considerably. By providing jobs to women, it contributes to poverty reduction—not only by providing jobs but also by ensuring fair wages and a safe working environment for their people which can help lift workers and their families out of poverty (Schoenmaker, 2020).

3.2 SDG 5: Gender Equality

Most of the workers in the garment industry are women. Throughout the garment industry, they are ensuring gender equality by maintaining equal pay for both men and women in the same positions, providing leadership roles, and providing a safe working environment against discrimination and women harassment, thereby contributing to gender equality in their workplaces (Gunatilaka, 2019).

3.3 SDG 8: Decent Work and Economic Growth

The apparel and textile industries are one of the most job-creating industries that contribute to the economic growth of the country. They also provide a better work environment to their workers.

3.4 SDG 12: Responsible Consumption and Production

On the growing concern of sustainability, the sourcing practices are also becoming more sustainable by using organic and recycled raw materials for production, thereby reducing the harmful environmental impact of apparel manufacturing processes. Utilising eco-friendly production processes, such as dyeing methods with minimal water usage or low-impact chemicals, helps mitigate the environmental footprint of the industry. The growth of sustainable fashion rather than fast fashion among young minds, which majorly focuses on the products of longevity, helps in improving responsible consumption and production (Radhakrishnan, 2020).

3.5 SDG 13: Climate Action

Adopting sustainable practices, such as reducing energy consumption in the processes, minimising greenhouse gas emissions in the processing and supply chain, and investing in renewable energy sources, contributes to climate action. Measures can be implemented to reduce carbon emissions throughout the supply chain, from production to transportation and distribution, to help mitigate climate change impacts (Moazzem et al., 2021).

3.6 SDG 14: Life Below Water

Taking measures to reduce water pollution from textile chemical processing processes, such as proper wastewater treatment and the use of eco-friendly chemicals, helps to protect the rivers and lakes around us.

3.7 SDG 15: Life on Land

Implementing land conservation measures to avoid landfills and promoting sustainable land management practices in the industry contribute to preserving the living organisms on land including humans (Arana et al., 2020).

3.8 SDG 17: Partnerships for the Goals

To promote the partnership goals, most industries are trying to collaborate with governments, NGOs, other industries, and stakeholders, which can facilitate knowledge sharing, resource mobilisation, and collective action towards achieving the SDGs. Also collaborating with suppliers, manufacturers, and retailers to promote sustainable practices throughout the supply chain fosters collective efforts towards the SDGs.

4 Current Scenario of the Sustainable Pillars

The textile and apparel industry is a highly labour-intensive supply chain that should establish a benchmark in social sustainability practices across all its sectors (Sudusinghe et al., 2018). The United Nations, which has introduced 17 SDGs, distinguishes the pillars of social, economic, and environmental development such as poverty eradication, clean water, sanitation, and so on. The current scenario of the sustainable pillars explains how they are balanced in the apparel industry. The social sustainability aspect of the pillars is about the social interaction within and among the stakeholders of the textile and apparel industry. Considering worker's health and well-being, many works have been conducted throughout the apparel industry.

A study shows that specific interventions should be done by the apparel manufacturing lines and global brands to improve the working conditions and well-being of the employees and labourers (Weziak-Bialowolska et al., 2019). Normally, these industries have been entitled to a stereotype as a 'traditional and old fashioned' industry, but improvement works have been done to equip the manpower with computational thinking ability for applying more systematic and innovative approaches towards apparel quality problems, thereby leading to a significant increase in productivity of an apparel or textile firm (Chong, 2019).

Corporate social responsibility (CSR) is a business model that focuses on the firm's work on its operation in ways to improve its societal aspects. The social sustainability dimension of the pillar is well balanced in terms of planning and implementation which could primarily help the companies to improve society and aid an optimistic image. It could also satisfy the social pillar of UN SDGs while contributing to societal goals. Environmental sustainability has played a great role of significance for the past few decades in the textiles and apparel industry. A considerable amount of methodologies have been applied to measure environmental sustainability in enterprises for environmental management of the company (Luo et al., 2021).

From the literature, we have found that relatively environmental sustainability is more concentric than the other pillars. Being the second largest polluting industry of all global manufacturing industries, environmental sustainability should be more of a concern for textile and apparel industries but the social and economic aspects shouldn't be ignored. Within the organisation, there is always a misinterpretation that sustainability is just about the environment and eco-friendly products that they produce. Many initiatives are taken based on preserving the environment from the harmful consequences of manufacturing. Product and process innovations are developed in terms of sustainability such as the development of the 3Rs, circularity which tells about cradle-to-cradle business models, closed-loop fashion, and sustainable designs that conserve natural resources environment, eliminate waste generation, and other eco-friendly benefits. The environmental pillar of UN SDGs is well taken care of according to the specifications of the UN by the T&A industry. T&A also plays a critical in taking urgent actions to combat the impacts of climate change and ensuring the elimination of those effects and consequences created by the production of such garments. Aligning the apparel and textile sector practices with the United Nations' Sustainable Development Goals is crucial for achieving a balanced approach to sustainability.

5 Importance of Balancing the Pillars

The three pillars of sustainability cannot be put into isolation as they are interconnected and interdependent. A small change in one of them will affect the other three in various ways.

5.1 Environmental Sustainability

5.1.1 Importance

The apparel sector is the major consumer of water, energy, and raw materials for their major processes. Sustainable practices help them in conserving these resources and reduce the industry's environmental footprint and carbon emissions. Garment production produces pollution in a large number of ways. These sustainable practices help mitigate water, air, and soil pollution, thus protecting the ecosystems and human health. This sector is a notable contributor to GHG emissions. By reducing these emissions, the world can attain climatic stability.

5.1.2 Implementation

Using eco-friendly raw materials such as sustainable cotton, hemp, bamboo, and recycled fibres which can be reused helps to reduce the environmental impact of raw material extraction and other processing. Doing the conventional practices effectively and efficiently by adopting technologies and practices that can reduce water and energy consumption, such as low water dyeing techniques and renewable energy sources in manufacturing, improves efficient manufacturing. Also, in the view of promoting the circular economy, implementation of waste management systems by recycling and upcycling initiatives can reduce the landfills and dumps (Radhakrishnan, 2020).

5.2 *Social Sustainability*

5.2.1 Importance

The apparel industry has been known for labour exploitation. So now to promote the SDGs, companies are trying to ensure fair wages, safe working conditions, and ethical labour practices to increase the industry's reputation and the well-being of workers (Gunatilaka, 2019). They are encouraging and interacting with the working community people to improve the stability and goodwill among them to contribute to an equitable society. Nowadays, all the consumers are increasingly demanding transparency in the whole supply chain process and ethical practices in production. Brands that stick to high social standards build stronger trust and loyalty among their customers (Kumar et al., 2017).

5.2.2 Implementation

Implementation and enforcement of fair wage policies, ensuring safe working environments through regular audits, and adherence to international labour standards can help in promoting social sustainability. It can also be improved by providing proper education and training opportunities for workers to improve their skills, leading to better job performance and career advancement (Gammanpila & Kodisinghe, 2020).

5.3 *Economic Sustainability*

5.3.1 Importance

Following sustainable economic practices ensures the long-term profitability and viability of businesses in the apparel sector. Investing more in sustainable practices rather than conventional methods encourages innovation, which leads to the development of new products and materials that can differentiate a brand in a competitive market. Now most consumers and retailers prefer products that are ethically made and environmentally friendly that do not cause any harmful effect on the environment.

5.3.2 Implementation

Adopting circular economy models where products are designed for durability, repairability, and recyclability, which extends their lifecycle, helps in promoting sustainable business models. Proper supply chain management helps to reduce waste, and improve logistics, thereby helping in cost-cutting (Kumar et al., 2017).

Investment in advanced manufacturing technologies such as 3D printing and automation helps in more effective and economic production. Balancing the pillars of the UN SDGs in the apparel and textile sector is essential for sustainable development. By integrating economic, social, and environmental sustainability, companies can ensure their long-term viability, contribute positively to society, and minimise their environmental impact. This holistic approach not only meets the growing demands of consumers and regulators but also positions the sector as a leader in sustainable practices, driving global progress towards achieving the SDGs.

6 Shareholders' Sustainable Development Plans

As far as sustainability is concerned, there is a huge amount of expectations on any organisation or company, not only in terms of sustainable decisions being made but also in its extent of activities to pursue and efficiently promote sustainable outcomes. With the ever-growing awareness of sustainability, today, society has become more considerate of environmental, social, and governance (ESG). Moreover, many investors consider these ESG factors while making their investments in the company. Also, purchase decisions made by consumers, especially the younger generations, exhibit their crucial search for sustainable products. With the current society claiming sustainability more than anything, not all ensure that they produce exactly sustainable and eco-friendly products. It is not practically possible for everyone to achieve it. Only in the presence of a proper stakeholder engagement, the true essence of sustainable goals can be tasted. The eight brief ways, in which stakeholder engagement helps any organisation, are discussed in the following.

6.1 Clearly Defining Sustainability for Every New Project

The idea of sustainability can be perceived in various ways, which vary from person to person. However, on a basic level, it can be based on things discussed below:

- A proper balance of economic, social, and environmental needs can refer to sustainability.
- Improvising and maintaining the overall availability of resources or the environmental conditions can also be called sustainability.
- Sustainability is what ensures the basic viability and long-term certainty of communities and business practices.

And many more influential definitions can be given. However, what sustainability means to a particular organisation or a particular project of the organisation is what is to be chosen wisely due to its crucial coherence required to achieve the desired outcomes. A proper stakeholder engagement does this job perfectly. It guides the organisational practices in line and also helps the stakeholders define sustainability precisely to their needs.

6.2 Receiving Feedback to Ensure Relevance

Sustainability can be brought about in numerous different aspects of an organisation. On a very basic level, these may include things like:

- Employee protection and welfare
- Water management systems
- Sustainably enhanced supply chains
- Improved working conditions
- Recycling and upcycling

And a lot more. Again, proper stakeholder engagement allows and guides the organisation well enough to identify the major issues that have to be solved as per the concerns of the shareholders. To do this in a much better way, questionnaires and surveys can be prepared to obtain their feedback in every meeting and interview. Opting for an area of sustainability that in turn would enhance stakeholder engagement can also be a wise choice.

6.3 Assumptions Made Should Be Checked Properly

These assumptions majorly include considering that the sustainability goals and measures taken earlier in the past as a constant thing that would be applicable for the future as well. However, these assumptions fail to be constructive in the growth of the organisation. Hence, there is a need for going through these assumptions and checking them, by engaging well with the stakeholders. This will help the organisation understand the factors that are unique and applicable to them, and make decisions that are more socially, economically, and environmentally sustainable.

6.4 Try Including More Perspectives

The stakeholder engagements inform the actions and decisions by involving all the key shareholders and stakeholder groups that are significantly impacted by the projects or the organisation. Industry and trade associations, suppliers, employees, communities, government, consumers, NGOs, environmental groups, regulatory groups, and agencies all constitute the stakeholders.

The much broader range of inputs that could probably be obtained through these engagements include:

- Possible issues and risks that could hinder the projects' success
- Opting for more sustainable alternatives
- Negative impacts that would probably affect the environment or society
- Considerable cultural issues that can emerge

These are likely to increase the long-term applicability of implementing sustainability measures.

6.5 Constantly Restoring Competing Views

With the common presence of multiple stakeholders, various opinions and expectations associated with it come into play. These might be on what sustainability is, in their opinions, and also the issues that they prefer to be considered and solved. However, proper stakeholder engagement will provide a platform for reconciling those competing opinions and views. Hence, by correctly identifying the right stakeholders and by mapping and matching their interests and influence, different stakeholder groups can be understood better and also it can ensure that the right people are consulted for each specific matter.

6.6 Building More Accountability

Having more accountability is a prerequisite for carrying out sustainable business practices. It's practically impossible to go and examine each organisation if it is sticking to sustainability as promised by them. So there should be a sense of self-awareness and its associated accountability that has to be built by every organisation. Stakeholder engagement turns out to help largely in achieving this as well. A few major ways in which it helps include:

- Collecting more and more data on different outcomes, feedback, and other engagement activities.
- To make better-informed decisions, demos on how the organisation is engaged can be created and presented.
- To keenly follow up on how their input impacted their outcomes, stakeholders should be kept informed regarding the regular reports and files.

More accountabilities can be brought about through regular and transparent reporting that maintains and enhances the capabilities of the organisation to its fullest. This eventually results in increased support, license, and trust to operate. This in turn will lead to the success of the project.

6.7 Rising Legitimacy

The importance of these stakeholder engagements is not only highlighted through the benefits they provide us with but also on a legal notion, and the necessity of these is also spoken about in frameworks like ISO 26000: Social Responsibility. Through proper engagement with the stakeholders, a company will be able to meet all the external requirements that would eventually raise the legitimacy desired to be achieved by the organisation.

6.8 Building Long-Lasting Relationships

Even from corporate and operational perspectives, the concept and ideology of sustainability are worth concerned about. The problem-free operation of the organisation in the long term is again correlated with stakeholder engagement. By building such long-lasting relationships with the stakeholders, the following benefits can be observed:

- Reduced risks to the organisation
- Changing with the trend and adapting to new, modern changes observed in the operational environment
- Quickly identifying all the areas of improvement (Hendricks, 2023).

6.9 Major Groups and Other Stakeholders

According to the 2030 Agenda, there were various intense consultations done and engagements practiced with different stakeholders and society; the major concerns were about the people who are more vulnerable and poorer. It is also found that in the upcoming decade, delivering these SDGs will pose serious requirements on a sense of urgency, ambitiousness, and decision-making ability.

Especially the *Decade of Action for Sustainable Development Goals* will need an immense number of quicker decisions and active cooperation and mobility of everyone, everywhere. This would transform the ever-flowing new ideas into concrete implementable solutions. To ensure proper implementation and following up of the 2030 Agenda, the *Division of SDGs* has been developed that, to a large extent, encourages and supports the involvement of stakeholders from completely differing sectors on a Global Forum. This DSDG also turns out to be helpful to national governments and stakeholders in enhancing their capacity-building opportunities by developing various research materials and publications. It also supports the proper exchange of knowledge and strengthens stakeholder engagement to a large extent.

However, the integral role played by the Major Groups and other Stakeholders (MGoS) to successfully adopt and implement the 2030 Agenda for Sustainable

Development is considerably significant. These MGoS are found to be so proactive in their proper implementation, initiatives, knowledge sharing, practicing, and frequent monitoring (UN, 2024). Governments and other people belonging to different sectors are said to partner up with them to work towards achieving the desired sustainability. Their participation in much higher-level intergovernmental activities and political forums is largely supported and taken care of by the UN DESA Office of Intergovernmental Support and Coordination for Sustainable Development (OISC).

6.10 Stakeholders Analysis Method for Sustainable Fashion Supply Chain

Based on studies conducted on novel neutrosophic approaches toward a Sustainable Fashion Supply Chain (SFSC), the stakeholders' roles are highlighted and emphasised. It tells about the Method of ACTors, Objectives, and strength Reports (MACTOR) approach. It deals with all the indeterminate and uncertain opinions of those who are supposed to make decisions. This research emphasises the fact that, despite the manufacturer playing the key role in producing the main product, they lack the supreme power to conduct a business because the consumer, fashion firms, and the government also play significant roles in shaping the fashion industry. Moreover, it is also said that the package providers, who produce eco-friendly products, are not given the right importance in terms of sustainability and its associated decision-making (Karadayi-Usta, 2023). Further researches highlighted the importance and support of various other people and stakeholders in achieving the desired goals not only UN but to adhere to the ideology of sustainability in general as well.

7 Strategies for Maintaining the Balance

The balance can be maintained by using some of the below-mentioned strategies.

7.1 Environmental Sustainability Strategies

- Usage of organic fibres such as cotton, hemp, bamboo, and recycled polyester that do not have any harmful impact on the environment. Obtain certifications like GOTS and OEKO-TEX to ensure that materials meet stringent environmental and social criteria.

- Implement technologies such as closed-loop water systems and water recycling to reduce consumption and pollution. Use energy-efficient lighting, heating, and cooling systems, and switch to renewable energy sources like solar and wind power.
- Use non-toxic dyes and chemicals and establish strict protocols for chemical handling and disposal to minimise environmental impact. Adopt zero-waste production techniques, such as fabric-cutting software that optimises patterns to reduce offcuts and implement comprehensive recycling programmes.
- Life cycle assessment of products can be done to evaluate and minimise the environmental impacts of products throughout their life cycle. Set and track targets for reducing greenhouse gas emissions, incorporating strategies like energy efficiency improvements and switching to renewable energy sources.

7.2 Social Sustainability Strategies

- Fair wage policies can be implemented to ensure all workers earn enough to meet their basic needs, supported by periodic wage reviews and adjustments. Ensuring all work environments can comply with safety regulations by implementing regular inspections and the provision of necessary safety gear and training.
- Support the right of workers to form unions and engage in collective bargaining to negotiate better wages and working conditions. Provide continuous education and skill development programmes to enhance workers' capabilities and career prospects.
- Invest in community initiatives such as healthcare, education, and infrastructure to improve the quality of life for local populations. Involve local communities in decision-making processes related to factory operations and sustainability initiatives.
- Conduct regular social audits to ensure compliance with ethical labour standards and address any issues promptly. Publish detailed sustainability reports that highlight social impacts, initiatives, and progress, fostering transparency and accountability.

7.3 Economic Sustainability Strategies

- Design products for longevity and reusability. This includes using durable materials and creating products that can be easily disassembled for recycling and implementing take-back schemes where customers can return used products. These items can be recycled into new products, reducing waste and resource consumption.
- Use technologies such as blockchain for traceability and transparency, which helps ensure that the materials are sourced effectively and that supply chains are

efficient. Reduce transportation costs and carbon emissions by sourcing materials from local suppliers, supporting local economies, and ensuring fresher, faster deliveries.
- Invest in R&D to discover sustainable materials (e.g., biodegradable fabrics, plant-based dyes) and innovative production methods (e.g., 3D knitting, which reduces fabric waste). Partner with technology firms, research institutions, and other stakeholders to leverage new technologies and sustainable practices.
- Adopt lean manufacturing techniques to minimise waste and improve efficiencies, such as just-in-time production and waste minimisation practices. Implement energy management systems to monitor and reduce energy consumption, including the use of energy-efficient machinery and renewable energy sources.

7.4 Integrated Strategies

Integrated strategies that can be adopted to promote are as follows:

1. Sustainable governance.
2. Stakeholder engagement.
3. Continuous improvement and adaptation.
4. Education and awareness.

Balancing the pillars of the UN SDGs in the apparel and textile sectors involves a strategic and integrated approach. By adopting circular economy models, optimising supply chains, investing in sustainable innovation, and maintaining transparency, companies can achieve economic sustainability. Fair labour practices, worker empowerment, community engagement, and transparent reporting are key to social sustainability. Resource efficiency, pollution reduction, and continuous environmental impact monitoring ensure environmental sustainability. Integrating these strategies ensures long-term viability, responsible growth, and a positive impact on society and the planet, aligning with the broader goals of the UN SDGs.

8 Future Scope of Research in Sustainability

The ever-growing need for sustainability and sustainable products, its booming awareness among people in society, and most importantly the immense concern that these sustainable practices have towards the environment, society, and the economy have not only attracted people in general but also highly knowledgeable professionals who in turn have made significant opinions in a global forum such as United Nations. The UN has a broader view towards developing and enhancing not only the present lives of people but their future too. Having clearly in mind the fact of the immense importance of the need for sustainable practices, they created the 2030 Agenda in the year 2015, which was 15 years ahead of the outcome. It not only included the major pillars of sustainability but also highlighted the importance of their correlation to benefit to the fullest. It talks about various factors and terms that pave the way towards achieving sustainability. This applies to all the sectors. With more highlights towards the fashion and textile sectors, sustainability has become a prerequisite to attract not only consumers but also the employees and shareholders. Moreover, in most cases, proper stakeholder engagement was what was found to be more helpful. Various research has been conducted emphasising the cooperation of everyone associated with the business to reach the level of sustainability target and thus adhere to the UN SDGs. However, despite the concept of sustainability gaining popularity, the practical implementations associated with it are not as expected. Still, there exists cases in which people don't choose the right perspective of sustainability that best suits their organisation or their projects. This is majorly due to the major mentality and thoughts that procure so shallow knowledge on this concept. Furthermore, awareness, deep understanding, and considering different perspectives can reflect positively. There can be various strategies adopted to achieve this as mentioned above. Moreover, there is still scope for various research to explore the less spoken aspects and factors that directly or indirectly influence the process of achieving sustainability. For example, throughout the whole time, discussions were around the single concept of organisations and their strategies; however, the social responsibilities of the public and the consumers of different products were undiscussed. But the fact is, they are the ones who make the society on a much larger scale. So, there are various such areas providing scope for further research.

9 Conclusion

The sustainable development goals made by the UN aim to transform the world on a very basic level. This option of transformation is brought to action only when precise sustainable practices are taken by the stakeholders of the apparel and textile sectors. In seeking to clarify the balance between the pillars of sustainability in terms of apparel and textile industries' contribution towards UN sustainable development goals, we have discussed the brief description of every UN SDG and what category of pillar of sustainability it falls on. There were a lot of development plans and primary contributions done by the apparel and textile firms to contribute to UN SDGs and society. The prior research in this area was only focused on a particular pillar. There wasn't a holistic approach that talked about the balance of each pillar in terms of stakeholders' behaviours. This study set out to examine the effects of the three aspects of sustainability on each other as they are interdependent. However, this study is limited based on its reliance on self-reported data, which is probably subject to bias. Future research should consider longitudinal studies to better understand the correlation between specific internal goals and specific SDG sub-targets. Additionally, qualitative research could provide deeper insights into the standardisation of the methodology to evaluate whether the industry is following the SDGs rather than looking up the analytical reports. While sustainability is an integral part of modern industrialisation, its potential effects and balance on other aspects of business cannot be overlooked. By addressing this issue, stakeholders and policymakers can help students achieve better sustainable outcomes while still benefiting from the positive aspects of UN SDGs. In conclusion, the apparel and textile sectors have made strides toward aligning with the SDGs, but significant work remains. By embracing sustainable practices and fostering collaboration across all stakeholders, the sector can make substantial contributions to a more sustainable and equitable global economy. This alignment not only supports the achievement of the SDGs but also ensures the long-term resilience and viability of the industry itself.

References

Arana, C., Franco, I. B., Joshi, A., & Sedhai, J. (2020). SDG 15 life on land: A review of sustainable fashion design processes: Upcycling waste organic yarns. In *Actioning the global goals for local impact: Towards sustainability science, policy, education and practice* (pp. 247–264).

Bain, P. G., Kroonenberg, P. M., Johansson, L. O., Milfont, T. L., Crimston, C. R., Kurz, T., Bushina, E., Calligaro, C., Demarque, C., Guan, Y., & Park, J. (2019). Public views of the sustainable development goals across countries. *Nature Sustainability, 2*(9), 819–825.

Boussemart, J. P., Leleu, H., Shen, Z., & Valdmanis, V. (2020). Performance analysis for three pillars of sustainability. *Journal of Productivity Analysis, 53*, 305–320.

Chong, B. (2019). Transforming the quality of workforce in the textile and apparel industry through digital empowerment. *The Educational Review, USA, 3*(8), 96–105.

Dalampira, E. S., & Nastis, S. A. (2020). Back to the future: Simplifying sustainable development goals based on three pillars of sustainability. *International Journal of Sustainable Agricultural Management and Informatics, 6*(3), 226–240.

Davidson, M. (2010). Social sustainability and the city. *Geography Compass, 4*(7), 872–880.

Duic, N., Urbaniec, K., & Huisingh, D. (2015). Components and structures of the pillars of sustainability. *Journal of Cleaner Production, 88*, 1–12.

Gammanpila, D. D., & Kodisinghe, K. A. M. S. (2020). The impact of training and development on employee performance of apparel industry in Sri Lanka. *Wayamba Journal of Management, 11*(2), 150–163.

Gunatilaka, H. (2019). *A literature review on gender and manufacturing: With special reference to employees in global garment factories in Sri Lanka, International Journal of Science and Research, 8*, 160–165.

Hendricks, A. (2023). *Advanced stakeholder management* [Online]. Available from https://simplystakeholders.com/stakeholder-engagement-sustainability/. Accessed 20th June 2024.

Karadayi-Usta, S. (2023). A novel neutrosophical approach in stakeholder analysis for sustainable fashion supply chains. *Journal of Fashion Marketing and Management: An International Journal, 27*(2), 370–394.

Kumar, V., Agrawal, T. K., Wang, L., & Chen, Y. (2017). Contribution of traceability towards attaining sustainability in the textile sector. *Textiles and Clothing Sustainability, 3*, 1–10.

Luo, Y., Song, K., Ding, X., & Wu, X. (2021). Environmental sustainability of textiles and apparel: A review of evaluation methods. *Environmental Impact Assessment Review, 86*, 106497.

Masterclass. (2022). *Economic Sustainability: Definition, Importance, and Examples by MasterClass* [Online]. Available from https://www.masterclass.com/articles/economic-sustainability. Accessed 20th June 2024.

Moazzem, S., Crossin, E., Daver, F., & Wang, L. (2021). Environmental impact of apparel supply chain and textile products. In *Environment, Development and Sustainability, Springer Nature* (pp. 1–19).

Purvis, B., Mao, Y., & Robinson, D. (2019). Three pillars of sustainability: In search of conceptual origins. *Sustainability Science, 14*, 681–695.
Radhakrishnan, S. (2020). Sustainable consumption and production patterns in fashion. In *The UN sustainable development goals for the textile and fashion industry* (pp. 59–75). Springer.
Schoenmaker, D. (2020). *The impact economy: Balancing profit and impact.* Working Paper, 4.
Stren, R., & Polèse, M. (2000). Understanding the new sociocultural dynamics of cities: Comparative urban policy in a global context. In *The social sustainability of cities: Diversity and the management of change, University of Toronto press* (pp. 3–38).
Sudusinghe, J. I., Jayaratne, R. P., & Kumarage, A. S. (2018, July). UN SDGs shaping sustainable supply chains: The case of apparel manufacturers in developing countries. In *2018 IEEE international conference on service operations and logistics, and informatics (SOLI)* (pp. 102–107). IEEE.
UN. (2012). *Report of the United Nations conference on sustainable development (A/CONF.216/16).* United Nations.
UN. (2024). *Stakeholder engagement by the Department of Economic and Social Affairs* [Online]. Available from *https://sdgs.un.org/stakeholders*. Accessed 20th June 2024.
Węziak-Białowolska, D., Białowolski, P., & McNeely, E. (2019). Worker's well-being. Evidence from the apparel industry in Mexico. *Intelligent Buildings International, 11*(3–4), 158–177.

Index

S. S. Muthu (ed.), *Sustainable Textile and Apparel Chain Management*, Sustainable Textiles: Production, Processing, Manufacturing & Chemistry,
https://doi.org/10.1007/978-3-031-80240-9

The manufacturer's authorised representative in the EU is Springer
Nature Customer Service Centre GmbH, Europaplatz 3, 69115 Heidelberg,
Germany. If you have any concerns regarding our products, please
contact ProductSafety@springernature.com

Printed and bound by CPI Group (UK) Ltd, Croydon, CR0 4YY
01/10/2026
02225969-0007